How Africa Eats
Trade, Food Security and Climate Risks

Edited by David Luke

LSE Press

Published by
LSE Press
10 Portugal Street
London WC2A 2HD
https://press.lse.ac.uk

Text © The Authors 2025

First published 2025

Cover design by Diana Jarvis

Print and digital versions typeset by Siliconchips Services Ltd.

ISBN (Paperback): 978-1-911712-34-3
ISBN (PDF): 978-1-911712-35-0
ISBN (EPUB): 978-1-911712-36-7
ISBN (Mobi): 978-1-911712-37-4

DOI: https://doi.org/10.31389/lsepress.hae

The full text of this book has been peer-reviewed to ensure high academic standards. For our full publishing ethics policies, see https://press.lse.ac.uk

Suggested citation:
Luke, David (ed) (2025) *How Africa Eats: Trade, Food Security and Climate Risks*, London: LSE Press. https://doi.org/10.31389/lsepress.hae
License: CC-BY-NC 4.0

To read the free, open access version of this book online, visit https://doi.org/10.31389/lsepress.hae or scan this QR code with your mobile device:

Contents

List of figures, tables and boxes

Figures

Tables

Boxes

Editors and contributors

Editor

David Luke is professor in practice and strategic director at the London School of Economics' Firoz Lalji Institute for Africa, where he oversees a programme on African trade policy. He is a former director of the African Trade Policy Centre at the UN Economic Commission for Africa (ECA), where he led the technical work on the protocols that make up the African Continental Free Trade Area (AfCFTA) Agreement. His research interests include boosting intra-African trade; the AfCFTA initiative; Africa's multilateral and bilateral trade relationships; and how trade policy intersects with industrialisation, structural transformation, inclusion, gender, public health and climate change. He is a member of the Board of TradeMark Africa and of the Council of the Africa Trade Foundation.

Contributors

Vinaye Dey Ancharaz is an independent consultant on trade and development. He was a senior lecturer at the University of Mauritius and has worked at the International Centre for Trade and Sustainable Development in Geneva and the African Development Bank. He has consulted with several organisations, including the World Bank, the African Development Bank, the United Nations, the Commonwealth Secretariat and the Indian Ocean Commission. He holds a first-class BSc (Hons) in economics from the University of Mauritius, MSc and PhD degrees in international economics from Brandeis University (USA) and an MBA from Imperial College London. He is a fellow of the Chartered Management Institute in the UK.

William Davis is an economist who has worked on international trade for over 15 years mainly focused on Africa, with the UK government, United Nations and various think tanks. An Oxford University graduate, he has over 30 publications in international trade. He is currently a visiting fellow at the LSE Firoz Lalji Institute for Africa.

Jamie MacLeod is an economics affairs officer at the United Nations Office of the High Representative for Least Developed Countries, Landlocked Developing Countries, and Small Island States in New York and a former policy fellow at the LSE Firoz Lalji Institute for Africa. His career spans more than a decade focusing on trade policies and negotiations, with a particular interest on the African continent, including at the UN Economic Commission for Africa, the International Trade Centre, the World Bank and the Ghana Ministry of Trade and Industry.

Olawale Ogunkola is professor of economics at the University of Ibadan, Ibadan, Nigeria. He has published and consulted extensively on trade and regional integration in Africa. He has been a visiting scholar at the International Monetary Fund, a visiting lecturer at the National University of Lesotho, a resource person at the World Trade Organization's regional trade policy course and a visiting professor at LSE.

Colette Van der Ven is an international trade lawyer specialising in sustainable development. As founding director of TULIP Consulting, Colette advises the public sector on legal and policy issues at the trade–environment–development nexus. She is also a visiting lecturer in international law at the Graduate Institute in Geneva. Previously, Colette worked as an international trade lawyer at Sidley Austin's international dispute settlement practice. She holds a juris doctor from Harvard Law School, and a master's in public policy from the Harvard Kennedy School of Government.

Acronyms and abbreviations

ACP	African, Caribbean and Pacific group of countries
AfCFTA	African Continental Free Trade Area
AfDB	African Development Bank
AFOLU	Agriculture, forestry and other land use
AGOA	African Growth and Opportunity Act (United States)
AGRA	Alliance for a Green Revolution in Africa
AMBs	Agricultural marketing boards
AMS	Aggregate measure of support
AMU	Arab Maghreb Union
ANRC	African Natural Resources Centre
AoA	The World Trade Organization's Agreement on Agriculture
AU	African Union
AU-IBAR	African Union – International Bureau for Animal Resources
BACI	Base pour l'Analyse du Commerce International (Database for the Analysis of International Trade)
BEC	Broad economic categories
BR	Biennial review under CAADP
BR1	First biennial review under CAADP
BR2	Second biennial review under CAADP
BR3	Third biennial review under CAADP
BSGI	Black Sea Grain Initiative
CAADP	Comprehensive African Agriculture Development Programme
CBAM	Carbon Border Adjustment Mechanism
CEMAC	Communauté Economique et Monétaire de l'Afrique Centrale (Central African Economic and Monetary Community)
CEN-SAD	Community of Sahel–Saharan States
CEPII	Centre d'études prospectives et d'informations internationales (Centre for Prospective Studies and International Information)

COMESA	Common Market for Eastern and Southern Africa
COP	Conference of the Parties to the UN Framework Convention on Climate Change. Note – this acronym can be used to refer to other conferences of parties to other international agreements, but in this book it refers to the conference related to the UN Framework Convention on Climate Change
DFIs	Development finance institutions
DFQF	Duty-free, quota-free
DRC	Democratic Republic of the Congo
EAC	East African Community
ECA	Economic Commission for Africa of the United Nations
ECCAS	Economic Community of Central African States
ECOWAS	Economic Community of West African States
EGA	Environmental Goods Agreement
EM-DAT	Emergency Events Database
EU	European Union
FAO	Food and Agriculture Organization of the United Nations
FBTAMS	Final bound total aggregate measurement of support (under the WTO Agreement on Agriculture)
FEWS NET	Famine Early Warning Systems Network
FFPI	FAO Food Price Index
FSA	Fisheries Subsidies Agreement
FTA	Free trade area
G20	Group of 20 countries with the largest economies by GDP
G33	Coalition of countries in agriculture reform negotiations at the WTO that support a special safeguard mechanism, exemptions of certain products from tariff reductions, and other protections to promote food security
GATT	General Agreement on Tariffs and Trade
GCF	Green Climate Fund
GDP	Gross domestic product
GHG	Greenhouse gas
GIs	Geographical indications (see glossary)
GPS	Global positioning system
GSP	Generalised System of Preferences
GTAP	Global Trade Analysis Project
ha	Hectare(s)

HS	Harmonised System. A system for classifying goods that are traded internationally into product categories.
ICBT	Informal cross-border trade
IFAD	International Fund for Agricultural Development
IFPRI	International Food Policy Research Institute
IGAD	Intergovernmental Authority on Development
IISD	International Institute for Sustainable Development
IITA	International Institute for Tropical Agriculture
IMPACT	International Model for Policy Analysis of Agricultural Commodities and Trade
IPCC	Intergovernmental Panel on Climate Change
IPRs	Intellectual property rights
ITC	International Trade Centre
IUU fishing	Illegal, unregulated and unreported fishing
LDCs	Least-developed countries
LSE	London School of Economics and Political Science
MFI	More Food International
MFIs	Microfinance institutions
MFN	Most-favoured nation (see glossary)
MNC	Multinational corporation
MSMEs	Micro, small and medium-sized enterprises
MT	metric tonnes, i.e. one thousand kilograms
NAIPs	National Agriculture (and Food Security) Investment Programmes
NDCs	Nationally determined contributions to addressing climate change under the Paris Agreement on Climate Change
NEPAD	New Partnership for Africa's Development
NERICA	New Rice for Africa
NFIDCs	Net food-importing developing countries
NGOs	Non-governmental organisations
NTBs	Non-tariff barriers
RAIPs	Regional agriculture investment programmes
RCP	Representative Concentration Pathway. This is the name given to climate change scenarios modelled by the IPCC
RECs	Regional economic communities
SACCOs	Saving and credit cooperatives

SACU	Southern African Customs Union
SADC	Southern African Development Community
SCCF	Special Climate Change Fund
SDT	Special and differential treatment for developing countries
SITC	Standard International Trade Classification
SPS	Sanitary and phytosanitary measures
SSG	Existing special agricultural safeguards under the World Trade Organization Agreement on Agriculture
SSM	Proposed new Special Safeguard Mechanism under the World Trade Organization Agreement on Agriculture
TBT	Technical barriers to trade
TESSD	Trade and Environmental Sustainability Structured Discussions at the World Trade Organization
TFTA	Tripartite Free Trade Area
tonne/ha	tonnes per hectare (i.e. tonnes of food produced per hectare of farmland)
TRIPS Agreement	Trade-Related Aspects of Intellectual Property Rights Agreement
TRQs	Tariff-rate quotas
UAE	United Arab Emirates
UK	United Kingdom
UMA	Union Maghreb Arabe (Arab Maghreb Union)
UN	United Nations
UN DESA	United Nations Department of Economic and Social Affairs
UNCTAD	UN Conference on Trade and Development
UNFCCC	UN Framework Convention on Climate Change
UNICEF	United Nations Children's Fund
US	United States
WAEMU	West African Economic and Monetary Union
WFP	World Food Programme
WHO	World Health Organization
WMO	World Meteorological Organization
WTO	World Trade Organization

Glossary

Ad valorem tariff A tariff where the tariff to be paid is determined as a percentage of the value of the goods being imported.

African Group at the WTO A group of African countries that are members of, or observers to, the World Trade Organization (WTO). In WTO negotiations, the group often speaks with one voice using a single coordinator or negotiating team.

African Growth and Opportunity Act A United States law that eliminates the tariffs due on imports from selected countries in sub-Saharan Africa for certain goods. The Act provides for more duty-free imports from these countries than from other countries that benefit from the United States' Generalised System of Preferences (see below).

Agadir Agreement A free trade agreement between Egypt, Jordan, Morocco and Tunisia.

Agricultural extension services Support provided to farmers to help them improve agricultural practices, productivity and sustainability including advice, information, equipment and other farming.

Aid for trade A subset of official development assistance focused on promoting and supporting international trade.

Amber Box Under the World Trade Organization's Agreement on Agriculture, the Amber Box refers to measures to support a country's agricultural industry that most distort trade and are subject to restrictions and reduction targets under the agreement. Examples of these subsides include price support regimes that regulate prices and production amounts; systems or targets for minimum prices for agricultural commodities, and highly subsidised insurance schemes and other forms of protection for farmers against low yields and/or price controls.

Bilateral trade (In this book) trade of African countries with non-African trading partners.

Black Sea Grain Initiative An initiative brokered by the United Nations following Russia's invasion of Ukraine. It allowed for exports of commercial food and fertiliser from three Ukrainian ports on the Black

Sea along an agreed route, with assurances from Russia that it will not attack vessels carrying this cargo along this route.

Blue Box Under the WTO's Agreement on Agriculture, the Blue Box refers to subsidies that may have some trade-distortive effects by limiting production or establishing production quotas, or payments to farmers for repurposing farmland. Blue Box subsidies are not counted towards a WTO Member's allowance for agricultural market support, which is limited under the Agreement. Blue Box subsidies are hardly used by developing countries as they involve direct payments which implies significant budgetary outlays. To date, no African country has made use of this type of subsidy.

Carbon Border Adjustment Mechanism (CBAM) A tax on certain imports in proportion to the greenhouse gases emitted during their production. The European Union currently has such a mechanism in force. For further details, see https://taxation-customs.ec.europa.eu/carbon-border-adjustment-mechanism_en.

Climate finance Loans, grants or other financing that helps countries to reduce ('mitigate') their emissions of greenhouse gases and/or adapt to the changing climate.

Comparative advantage A country has comparative advantage in a particular product or service when it has a lower opportunity cost of producing and exporting that product or service to other countries and thereby enhancing trade between countries.

Competition policy Policies to ensure that companies compete with each other to provide the best value for money to their customers and/or suppliers, instead of either colluding with each other to offer less favourable terms, or dominating a particular market and suppressing competition.

Computable general equilibrium model A computer-based model of the economy that allows users to estimate the effect of various economic or policy changes on the economy as a whole.

De minimis **domestic support** Minimal amounts of domestic support that, owing to their small size (and consequent minimal distortive effect), are allowed under WTO agreements even though they distort trade.

Derivatives Agreements between parties to buy or sell an asset, or make or receive payments, at a future date. Derivatives are based on the value of an underlying asset, such as a currency, index, interest rate or group of assets. They can be used to hedge against price movements, inter alia.

Development Box This is a group of subsidies that developing countries can provide without limitations under the World Trade Organization's

Agreement on Agriculture. These subsidies include inputs such as irrigation systems and fertilisers for low-income producers and outlays for the acquisition of machines and provided they are used to promote agricultural and rural development and form an integral part of development programmes.

Digital trade International trade that is enabled by information communication technology. It can include trade in goods and services and can take place in all sectors of the economy.

Duty-free, quota-free Without limit on the amount of trade that is traded without tariffs.

Elasticity The percentage by which one economic variable changes in response to a 1 per cent change in another economic variable.

Enteric methane Methane emitted by animals during their digestive process.

European Green Deal A range of policies pursued by the European Union (EU) to transition to no net greenhouse gas emissions by 2050, while decoupling economic growth from resource use. Measures included in the Green Deal include the RePower EU initiative (shifting to renewable energy), the EU Green Deal Industrial Plan (enhancing the competitiveness of industries that will help the continent to cut its greenhouse gas emissions) and ensuring access to affordable energy. As part of this, the European Union adopted the 'Fit for 55' package of measures that aim to reduce the bloc's greenhouse gas emissions by 55 per cent by 2030 (when compared to 1990 levels). It includes a target to increase the share of renewables in the EU's energy mix, and a commitment to raising the EU's internal carbon price.

Everything but Arms A European Union scheme that eliminates taxes and import quotas due on import from least-developed countries on all products except arms and ammunition.

Factors of production These are land, labour, capital (e.g. machinery, tools and buildings) and entrepreneurship (which combines the other factors in new ways).

Fit for 55 See European Green Deal above.

Food security When all people, at all times, have physical and economic access to sufficient, safe and nutritious food that meets their dietary needs and food preferences for an active and healthy life.

Foreign direct investment An investment from a company in one country into a business in another country.

Free-on-board price In international trade means the price of goods at export, excluding the costs of insurance and freight for travel to their destination.

Generalised system of preferences A greater degree of market access for goods provided by developed countries to developing countries that the former has selected.

Geographical indication A sign or name that identifies a product as having a specific geographical origin and characteristics that are related to that origin. GIs are used to protect products that have a reputation or quality that is due to their geographical origin.

Green revolution A great increase in production of food in developing countries, beginning in the mid-20th century, largely due to the introduction of new varieties of crops, mechanisation and other agricultural inputs.

Groupage The aggregation of informal trade wherein groups of traders would bring smaller individual assignments of goods together into a larger, consolidated consignment.

Harmonised System (HS) A system for classifying goods that are traded internationally into product categories, provided by the World Customs Organization.

Herbicide An agent for killing or inhibiting the growth of unwanted plants.

Informal cross-border trade Trade that crosses borders of adjacent countries that is carried out by traders who are unregistered with national tax authorities and business regulators. Typically it is not recorded in national statistical systems for measuring trade.

Intellectual property rights Legal rights that allow owners of intellectual property (e.g. product designs and artistic works) to determine who is legally allowed to use their intellectual property, and how. Examples include patents for inventions, copyright, trademarks and designs.

Intra-African trade Trade where both the exporting and importing country are African.

Low-/high-unit-value foods Foods that have a relatively low/high price per unit of weight/volume.

Malabo Declaration A declaration by heads of state and government of African Union member states adopting goals for agricultural production in Africa, to be achieved by 2025.

Metrology The science of measurement.

Most-favoured nation A common principle in trade agreements based on the idea that countries treat each other with no less 'favour' than they treat other countries in aspects such as tariffs on traded goods or conditions of access for service suppliers.

Most-favoured nation tariffs Tariffs that a country applies to imports from members of the WTO with which that country does not have a trading arrangement under which it has committed to tariffs lower than its commitments at the WTO.

Nominal figures Monetary values expressed using the value at the time that they relate to. These contrast with 'real' or 'constant-price' figures, which show trends in the real value of economic variables over time, stripping out the effect of inflation.

Non-communicable diseases Diseases that cannot be directly transmitted between people. These are due to a combination of genetic vulnerabilities and environmental factors, such as poverty, living in polluted surroundings, unreliable access to food or poor diet, tobacco, alcohol and other substance abuse.

Non-tariff barriers Restrictions to trade that do not take the form of a tariff, including quotas, embargoes, sanctions, customs and transit formalities, documentation or standards requirements.

Official development assistance Government aid that promotes and specifically targets economic development and welfare of developing countries.

Partial equilibrium model An economic model that estimates the impacts of policy changes on one or more markets of the economy but not the entire economy. Such models normally allow the modeller to look in more detail at the economic impacts on specific sectors of the economy, and to use more up-to-date information, than do general equilibrium models.

Pastoralist A farmer who breeds and takes care of animals.

Peace clause An agreement by WTO member states not to launch legal challenges under the WTO's dispute settlement procedures against food security programmes of developing countries that exceeded the levels of support to agriculture to which the country implementing the programme had agreed, provided that the programme met certain transparency conditions. In this book, we refer to the 'peace clause' agreed at the 2013 WTO Ministerial Conference and extended indefinitely in 2015. An earlier 'peace clause' had expired in 2004.

Peri-urban areas Non-urban areas immediately adjacent to cities or towns.

Public stockholding programmes Governments use public stockholding programmes to purchase, stockpile and distribute food when needed. Specifically, they provide (1) emergency stocks to reduce the vulnerability of consumers to supply disruptions or food price shocks in emergencies; (2) buffer stocks to stabilise prices within the domestic market to avoid excessive volatility; and (3) stocks for domestic food distribution or for external food aid.

Randomised controlled trial An experiment to measure the effect of a new measure, policy (or, in medicine, a treatment) in which participants are selected at random. This is often considered to be the most rigorous way to measure the effectiveness of different interventions.

Regional economic communities (RECs) The geographic groupings of African countries that form the building blocks for regional coordination within the African continent. The term often refers to the eight African Union-recognised RECs but can include other formations such as the Southern African Customs Union or the Indian Ocean Commission.

Rules of origin The criteria that a good needs to satisfy to be considered to 'originate' within a country, and therefore eligible to benefit from the trade preferences accorded to that country by partner countries.

Safe trade Sector-specific practices on issues such as border health screening, testing and certification, truck crew sizes, digitalised trade procedures, electronic cargo tracking and information sharing.

Sanitary and phytosanitary measures Measures to ensure that imported products uphold minimum standards of safety, to protect human, plant or animal life or health.

Schedule of concessions A document outlining how a country will reduce barriers to trade as part of a trade agreement

Sensitivity analysis Testing to see how much a model's results vary according to its assumptions. This can be used to determine how confident we can be that a model's results are accurate.

Smallholder Small-scale farmers, pastoralists, forest keepers, fishers who manage areas varying from less than one hectare to 10 hectares.

Special safeguard mechanism A proposed safeguard for developing countries to raise tariffs if imports of those products surge or prices decline.

Special safeguards Existing special agricultural safeguards under the World Trade Organization Agreement on Agriculture. For certain products in certain importing countries, these allow for the importing country to increase trade tariffs on agricultural imports if they rise above a certain level, without having to prove that they are causing serious harm to the importing country's domestic industry. For more information, see https://perma.cc/F7M6-C2KL

Staple foods The staple foods in a particular geographic location or community are those foods that people in that group consume regularly and that make up a significant part of their energy consumption.

Structural transformation A shift of economic activity between different economic sectors. It normally refers to shifts towards more productive sectors, with the result that economic output rises.

Supply-chain financing This involves a supplier receiving early payment of an invoice by a finance company. The business that has purchased the goods or service then pays the funder once the invoice is due.

Tariff-rate quotas These apply a lower tariff rate on imports of a given product within a specified quantity and requires a higher tariff rate on imports exceeding that quantity.

Tariffs Taxes on the importation of goods.

Technical barriers to trade Requirements for product characteristics, processes and production methods – combined with methods to assess whether traded products meet those requirements – that can impede trade.

Trade deficit The amount by which the value of a country's imports exceeds that of its exports. The agricultural trade deficit is the amount by which the value of its imports of agricultural goods exceeds the value of its exports of agricultural goods.

Trade facilitation initiatives Programmes or efforts made to expedite the movement, release and clearance of goods, including goods in transit.

Trade integration The process through which two or more states within a broadly defined geographic group reduce economic barriers to trade including tariffs, but also non-tariff issues such as the harmonisation of standards or customs coordination.

Trade preferences Where an importing country provides preferential treatment (e.g. lower tariffs, higher quotas) for imports from a particular exporting country, compared to imports from other countries.

Undernourishment Not eating enough food to continue to be in good health It is understood that the average person needs a minimum of 1830 kcal per day to avoid undernourishment.

Upstream Parts of value chains in which products are relatively unprocessed, or that concern inputs used in the production of other products. Processing occurs in downstream parts of the value chain.

Value chains The various stages in the production of goods and services. These are sometimes spread over several countries.

Preface and acknowledgements

The purpose of this book is to take a fresh look at why food deprivation persists at the scale it does in many parts of Africa and the policy questions this raises. We do this by focusing on food security at the intersection between trade, agricultural and climate policies. The high prevalence of hunger and undernourishment in many parts of Africa is quite rightly a matter of international concern. Frameworks established by the United Nations and stakeholders such as the Alliance for a Green Revolution in Africa (AGRA) and the International Food Policy Research Institute monitor the risks. The second sustainable development goal (SDG 2) affirms the global aspiration to end hunger and provides a basis for tracking progress.

This book builds upon these frameworks. It provides an original conceptual approach for depicting the relationship between food security, food trade and climate risks. It analyses production, consumption, policies, resources, finance, investment, institutions, actors and capacities. Within Africa, it follows formal and informal trade flows of the most widely consumed food products (or 'basic foods'): yams, cassava, rice, maize, wheat, meat, poultry and fish. It uncovers what impact on food security could be expected from implementation of the African Continental Free Trade Area at the intra-African level. It examines the trends in food trade between African and external partners including an appraisal of the effect of Russia's invasion of Ukraine on Africa's food security. It assesses whether World Trade Organization rules on agricultural trade are a stumbling block or a stepping stone to achieving food security in Africa.

The book follows *How Africa Trades* (LSE Press, 2023) as the second major research output of the Firoz Lalji Institute for Africa (FLIA) Africa Trade Policy Programme (ATPP) at the London School of Economics and Political Science. Through its research and impact activities, the ATPP is committed to three main objectives. First, it aspires to demystify African trade policies and propagate a deeper and broader understanding of how these and related policies – in this book, agricultural and climate policies – impact the lives of ordinary Africans and the continent's development aspirations. Second, it aims to provide up-to-date information that is easily reachable through open access publication on Africa's trade data, trade agreements and policies, with analysis to enhance clarity and guide meaningful interpretation. Third, it seeks to empower policymakers, stakeholders, scholars and others to interrogate the effectiveness of policy choices including the implementation dimensions from a normative perspective that is pro-development and inclusive and gives precedence to over-coming pervasive poverty on the continent. Readers are encouraged to use the book's insights – including what they find themselves in agreement or disagreement with – to engage on ending food deprivation in Africa.

Yet again, it has taken a village to produce a book. Resources and means to support various aspects and stages of the research were provided by the Friedrich Erbert Stiftung (FES) Geneva Office and the FLIA. Thanks are due to Hansjorg 'Hajo' Lanz, Renate Tenbusch, Yvonne Bartmann and Sabine Dorfler of the FES Geneva Office; and to FLIA colleagues, Tim Allen, Martha Geiger Mwenitete, Fadil Elobeid, Lesley Orero, Tosin Adebisi, Elinam Yebu, Mark Briggs, Sofija Spasenoska, Anna Williams and Eunice Hansen-Sackey for assorted engagements to enhance the ATPP's work. Ade Freeman of the FAO Africa Regional Office in Accra and Ify Ogo of UNDP provided valuable advice at the initial phase of the research.

The following attended virtually or in-person one or both of the workshops held at LSE in November 2022 and June 2023 to discuss research findings: Fousseini Traoré (Consortium of International Agricultural Research Centres, CGIAR), Gerald Masila (East African Grain Council), Suffyan Koroma (FAO), Calvin Maduna (Institute for Agriculture and Trade Policy), Antoine Bouët (International Food Policy Research Institute), Duncan Mac-Fadyen (Oppenheimer Generations Research and Conservation), Rendani Nenguda (Oppenheimer Generations Research and Conservation), Laetitia Pettinotti (ODI), Max Mendez-Parra (ODI), Facundo Calvo (International Institute for Sustainable Development), Poorva Karkare (European Centre for Development Policy Management), Hanne Knaepen (European Centre for Development Policy Management), Yvonne Bartmann (FES, Geneva), Lennart Oestergaard (FES, Berlin), Marjam Mayer (FES, Berlin), Edwini Kwame Kessie (WTO), Chibole Wakoli (WTO), Melaku Desta (Economic Commission for Africa), Simon Mevel (Economic Commission for Africa), George Kararach (African Development Bank), Kasper Vrojilk (University of Goettingen), Nick Westcott (School of Oriental and African Studies, SOAS), Molly Foster (University of Greenwich), Abbi Kedir (University of Sheffield), Andrew McKay (University of Sussex), Amir Lebdioui (Oxford University) Frank Lisk (University of Warwick), Elitsa Garnizova (LSE Trade Hub/LSE Consulting), Daniela Baeza Brein Bauber (LSE Research and Innovation), Elisabeth Robinson (LSE Grantham Institute on Climate Change and the Environment), Roberta Pierfederici (LSE Grantham Institute on Climate Change and the Environment) and Catherine Boone (LSE Firoz Lalji Institute for Africa Internal Board). While the research team benefitted from their insights, none is responsible for the issues covered in the book or the conclusions that are reached.

And, once again, it has been a pleasure to work with the superb team at LSE Press including Sarah Worthington (Chair), Alice Park (Managing Editor), Niamh Tumelty (Managing Director) and Elinor Potts (Communications Coordinator).

David Luke, *London, February 2025*

1. Introduction: towards a reassessment of food deprivation in Africa

David Luke

Why do images and reports of starving and malnourished Africans pop up so often in the media? What are the actual dimensions of the problem? What has trade and climate got to do with it? These are among the questions this book seeks to answer, in an effort to explain why Africa struggles with food availability and stability that are the essential pillars of food security, and what can be done about it. The intersection between trade and agriculture policies and a changing climate is fundamental to the enquiry.

The scale of food deprivation in Africa is sobering. The United Nations (UN) estimated that a fifth of the African population, some 280 million people, were undernourished in 2022. In the same year, even more people, 340 million Africans, a quarter of the population, lived with the uncertainty of access to food and sufficient consumption that is the day-to-day experience of severe food insecurity (FAO et al. 2023).

The book is appearing at a time of a surge in food prices that followed the Covid-19 pandemic and turbulence in global food markets. Adding fuel to the inflationary spiral was the war that started with Russia's invasion of Ukraine, two major suppliers to world food and fertiliser markets. The Food and Agriculture Organization of the United Nations (FAO) Food Price Index (FFPI) registered 159.7 points in March 2022, a few weeks after the war started. This was the highest value of the FFPI in 22 years, reaching well above earlier peaks during the 2007–2008 financial crisis and the 2011 commodity price surge (Shahbandeh 2024). The rising cost of food as well as increased frequency of extreme weather events that impact agricultural production has seen a tightening of export stocks against increased import demand. World Trade Organization (WTO) surveillance of Group of 20 (G20) economies that together account for 75 per cent of global trade revealed that these countries had 19 export restrictions on food, animal feed and fertilisers in place as of

---

May 2023 (WTO 2023). Since most African countries are net food importers, their access to food is largely dependent on global markets. The UN Conference on Trade and Development (UNCTAD) estimates that 82 per cent of African countries' basic food comes from outside the continent (UN Trade & Development UNCTAD n.d.). African households in the poorest countries are especially vulnerable to global price shocks and supply volatility.

Food insecurity in Africa is spreading in step with poverty, demographic and urbanisation trends. In 2015, 206 million Africans, or 17 per cent of the 1.2 billion population, were severely food-insecure. By 2022, this had increased to a quarter of the 1.4 billion population. While all of Africa's five regions – North, West, Central, Eastern and Southern – had more severely food-insecure people in 2022 than in 2015, West and Eastern Africa had the largest increases in the share of people affected. The number of severely food-insecure people in West Africa more than doubled between 2015 and 2022, from 41 to 95 million. In Eastern Africa, it increased by a quarter, from 87 to 132 million people (FAO et al. 2023). This partly reflects rising poverty rates and vulnerabilities to desertification in West Africa's Sahel and recurring droughts in the Horn and adjacent areas in Eastern Africa.

The headline data on severe food insecurity mirrors data on the prevalence of undernourishment as an indicator of hunger which in Africa as a whole has risen steadily since 2010. Africa has relatively high global shares of low birth weight, stunting and child wasting (a life-threatening condition caused by insufficient nutrient intake and poor nutrient absorption; affected children are dangerously thin, with weakened immunity and a higher risk of mortality). Child obesity is spreading as a mainly urban and peri-urban phenomenon (FAO et al. 2023). This is part of an emerging trend in which access, availability and consumption of highly processed foods in African urban settings is playing a part in the rise of non-communicable diseases (Malhotra and Vos 2021). This mirrors the global trend in the prevalence of non-communicable diseases that has been observed in middle and high-income countries (Kang, Kang and Lim 2021).

1.1 Defining, measuring and monitoring hunger

Several UN agencies, notably FAO, the International Fund for Agricultural Development (IFAD), the United Nations Children's Fund (UNICEF), the World Food Programme (WFP) and the World Health Organization (WHO), work together to systematically monitor progress towards ending hunger and the state of access of all people to safe, nutritious and sufficient food. These agencies are also part of the UN Committee on World Food Security that among other remits is tasked with quantifying and evaluating hunger and food security across the world. An important initiative spawned by the work of the UN agencies is the data-packed annual *State of Food Security and Nutrition in the World*. This exercise in tracking nutritional and food security targets is integral to the work of the UN on Sustainable Development Goal 2, which is concerned with eliminating global hunger.

Global monitoring of the status of nutrition and differentiating between levels of food security is erected on foundational clarity on the meaning of these concepts. In the UN system, food security is defined as when all people, at all times, have physical and economic access to sufficient, safe and nutritious food that meets their dietary needs and food preferences for an active and healthy life. This characterisation is built upon four dimensions that are integral to food security (Resnick and Swinnen 2023).

Physical *availability* of food is concerned with the 'supply side' of food security and is determined by the level of food production, stock levels and net trade. Since the various factors affecting food production and consumption are unevenly distributed across time and space, trade between countries and regions can help to adjust to changing conditions affecting food production including to climate change (FAO 2022). Economic and physical *access* to food is concerned with household level food security in relation to the role of incomes, expenditure, markets, prices, public and humanitarian support programmes in achieving food security objectives. Food *utilisation* or consumption is concerned with the sufficiency of energy and nutrient intake by individuals in relation to intra-household distribution of food, which combined with the biological utilisation of food consumption determines the nutritional status of individuals. And food *stability* is concerned with the stability of the other three dimensions.

1.2 The food system, challenges and global response

The determinants of nutritional status and food security are embedded in the food system that is in place. A food system is defined as the sum of policies, resources, actors, capacities and interactions along the food value chain – from input supply and production of crops, livestock, fish and other agricultural commodities to marketing, transportation, processing, wholesaling, retailing, preparation of foods, consumption and disposal (AGRA 2022). A 2021 UN summit on food systems recognised its broad impact on employment, incomes and development as a whole. Built into this understanding of the multifaceted impact of food systems is that enabling policy environments, cultural norms around food, and environmental sustainability are essential for the functioning of food systems. The International Food Policy Research Institute (IFPRI) estimates that food systems account for as much as 34 per cent of greenhouse gas emissions stemming from up- and downstream agriculture-related activities, with two-thirds of this arising from agriculture, forestry and other land use, or AFOLU (IFPRI 2022).

Africa's difficulties with undernourishment and hunger are related to challenges with the continent's food system. These challenges include but are not limited to: dominant smallholder farming practices, limited commercialised agriculture and low productivity; policy failure; resources, institutional and capacity gaps in agricultural markets; uneven penetration of technologies that enable agricultural productivity and engender a green revolution; distortions in global food markets underpinned by inequitable trade rules; climate vulnerabilities; and political instability, conflict and demographic trends.

To the extent that access to food is a universal human right, the UN plays a leading role in the global response to hunger everywhere. To this end, there are three organisations that lead the UN's work on food security, based together in Rome. The FAO is the UN specialised agency that leads international efforts to defeat hunger and improve nutrition and food security. IFAD was created as the UN's funding arm to mobilise investments in rural infrastructure and farm extension services, and build resilience against climate change across the developing world. WFP is leading the UN's humanitarian effort in delivering food aid to vulnerable communities, especially in conflict situations, climatic catastrophes and other emergencies. This includes its heroic role in making food available to vulnerable communities during the Covid-19 pandemic. The work of the WFP was recognised in the award of the 2020 Nobel Peace Prize. Prominently displayed at its headquarters in Rome is the citation that accompanied the prize, which says in part that the honour was given for its 'efforts to combat hunger, for its contribution to bettering conditions for peace in conflict-affected areas and for acting as a driving force in efforts to prevent the use of hunger as a weapon of war and conflict' (The Nobel Peace Prize 2020 n.d.). Today, UN member states put the WFP at the heart of global humanitarian response efforts: as of 2022, the WFP accounted for 24 per cent (about $12.2 billion) of the UN operational budget worldwide. The efforts of the FAO to align its technical support with African agricultural policies and of the WFP to source food aid from within Africa where possible to avoid undermining production systems are outlined in Chapter 4.

Another key organisation in the global response to tackle undernourishment and hunger is the Geneva-based WTO. Restrictions on international food trade can raise food prices and hit developing countries hardest. Subsidies in countries that can afford them contribute to global food availability but disincentivise production in poorer countries through price suppression. The WTO is where discussions take place to limit restrictions, subsidies and related complications, and to facilitate international food trade flows (Pangestu and Van Trotsenburg 2022). The role of the WTO in international food trade is discussed in Chapter 9.

1.3 Research focus and the book in outline

The first task of the research team that came together to examine the nexus between trade, food security and climate risks under the auspices of the Africa Trade Policy Programme at the London School of Economics and Political Science (LSE) Firoz Lalji Institute for Africa was to clarify *what* Africa eats, the basic foods with the highest contribution to calorie intake across the continent. It is understood that the average person needs a minimum of 1830 kcal per day to avoid undernourishment and 2360 kcal per day for optimal health (UN Trade & Development UNCTAD 2024).

As the discussion evolved, and following a thorough literature review, the research team investigated a number of further specific questions. How might a conceptual and quantitative approach be framed for depicting the relationship between food security, food trade and climate risks? What is the regional breakdown in the production and consumption of the food products identified as Africa's basic foods? How does Africa fare in terms of comparative yields and productivity with other parts of the world in the production of these food products and what are the climate risks related to their production? What is the implementation record of agricultural policies such as the Comprehensive African Agriculture Development Programme and the Malabo Declaration that were adopted within the framework of the African Union (AU) to boost production and productivity? How is implementation impacted by resources or lack of resources invested in African agriculture? How effective in terms of capacities are the various actors and institutions that operate within Africa's food system? How and where do the basic foods feature in intra-African trade flows? What is the likely impact of the African Continental Free Trade Area (AfCFTA) on intra-African food trade? Do the AfCFTA Agreement and Protocols contain specific provisions on agriculture? How could they be leveraged to enhance Africa's food security? If non-tariff barriers are more important than tariffs as obstacles to intra-African food trade, what should be done about them? To the extent that Africa as a whole is a net food importer, what are the main features of Africa's food trade with countries outside the continent? Which are these countries or bilateral food trade partners? The WTO is the global trade regulator – do its rules help or hinder the achievement of food security in Africa? These are the questions that occupied the research team. Over 18 months of research, two workshops were organised at LSE to review initial findings. Experts from the FAO, the WTO, CGIAR, IFPRI, the Institute for Agriculture and Trade Policy, the International Institute for Sustainable Development, the Economic Commission for Africa of the United Nations, the African Development Bank, Friedrich Ebert Stiftung, ODI Global and the LSE Grantham Research Institute on Climate Change and the Environment participated in these discussions, offering insights and advice.

The deconstruction of *how* Africa eats is the overriding objective of this book. Following this introductory chapter, the second chapter provides a conceptual approach for thinking about food trade, food security and climate risks. It uncovers exactly what Africa's food trade is, first by looking into Africa's agricultural exports and imports broadly before narrowing down to examine the trade flows at the product level driving such food trade. This is to help the reader understand exactly what is meant in this book by Africa's 'food trade', the consequences of this trade for food security and the attendant climate risks.

Chapter 3 focuses on what Africa eats and entails a review of eight key products that form the basket of foods that are essential for Africa's food security (wheat, yams, cassava, maize, rice, poultry, meat and fish), drawing upon FAO data on production and consumption, volume and value, with regional breakdowns. This is complemented by some illustrations of climate risks related to both emissions in production and the effect of changing weather patterns.

Chapter 4 assesses implementation of the AU-adopted agricultural policies in relation to resources from public budgets, private investment, foreign direct investment, foreign aid flows and climate finance for adaptation and mitigation in the agricultural sector. The capacities of various actors and institutions operating in Africa's food systems are also brought into focus. Chapters 5 and 6 analyse intra-African food trade flows, with the former focusing on recent trends and regional aspects of how the basic foods feature in intra-African trade flows. The latter undertakes a detailed partial equilibrium modelling exercise to assess the expected impact of the AfCFTA on intra-African food trade flows at the product level. Chapter 7 investigates the extent to which provisions in the AfCFTA Agreement and Protocols support agricultural development and the opportunities offered by this continental legal framework for advancing food trade and food security including through disciplines on non-tariff barriers. Chapter 8 reviews bilateral food trade flows (i.e. those between Africa and its non-African partners), their composition, the trade policies and some problematic issues that underpin these flows. Chapter 9 examines the WTO regulatory framework on agriculture and implications for food security in Africa. Relevant aspects of WTO climate and sustainability discussions are also brought into the focus. Chapter 10, the concluding chapter, highlights the main insights from the preceding chapters as a call to action by African policymakers, stakeholders and campaigners on ending hunger and development partners.

It is our expectation that the book will contribute to the current knowledge base on the policy landscape that impacts trade, food security and climate risks in Africa for more informed deliberations at various levels of policymaking, advocacy and scholarly and pedagogical pursuits

1.4 Open access publication

As was the case with the Africa Trade Policy Programme's previous book, *How Africa Trades* (2023), published by LSE Press, this book is being made available on an open access basis. All the datasets used in our analysis are, where possible, publicly available (not behind paywalls), with sources detailed in the reference sections at the end of each chapter. Where website addresses are liable to change, we have used Harvard's perma.cc resource to preserve online sources and ensure they are permanently available to readers. Aside from the inherent virtue of putting the result of social science research within the reach of any reader anywhere in the world, open access publication is especially beneficial to readers in Africa, where the relative cost of books and periodicals is high. Moreover, apart from the output of the UN agencies, the development banks and other international organisations, very little independent research is being carried out and published on Africa's food security in relation to trade and climate. Comments and feedback provided by readers are welcome and useful, and advance open social science. Please send this to Africa@lse.ac.uk. Engaging with the material covered in the book through posts on LinkedIn, X (formerly Twitter) (@AfricaAtLSE), Facebook and other social media is also welcome.

References

AGRA (2022) 'Africa Agriculture Status Report. Accelerating African Food Systems Transformation', Issue 10, Nairobi: Alliance for a Green Revolution in Africa. https://perma.cc/8KT3-MXY2

FAO (2022) *The State of Agricultural Commodity Markets 2022: The geography of food and agricultural trade: Policy approaches for sustainable development*. Rome, FAO.

FAO; IFAD; UNICEF; WFP and WHO (2023) 'The State of Food Security and Nutrition in the World 2023: Urbanization, Agrifood Systems Transformation and Healthy Diets across the Rural–Urban Continuum', FAO.

IFPRI (International Food Policy Research Institute) (2022) '2022 Global Food Policy Report: Climate Change and Food Systems', International Food Policy Research Institute (IFPRI). https://perma.cc/6XAN-YM2H

Kang, S Sooyoung; Kang, Minji; and Lim, Hyunjung (2021) 'Global and Regional Patterns in Noncommunicable Diseases and Dietary Factors across National Income Levels', *Nutrients*, vol. 13, no. 10. https://doi.org/10.3390/nu13103595

Malhotra, Swati; and Vos, Rob (2021) 'Africa's Processed Food Revolution and the Double Burden of Malnutrition', *IFPRI Blog*, 11 March. https://www.ifpri.org/blog/africas-processed-food-revolution-and-double-burden-malnutrition

The Nobel Peace Prize 2020 (n.d.) 'The Nobel Prize'. https://perma.cc/29WH-AS67

Pangestu, Mari Elka; and Van Trotsenburg, Axel (2022) 'Trade Restrictions Are Inflaming the Worst Food Crisis in a Decade', *Voices*, 6 July. https://perma.cc/348H-44QQ

Resnick, Danielle; and Swinnen, Johan (eds) (2023) *The Political Economy of Food System Transformation: Pathways to Progress in a Polarized World*, Oxford University Press. https://doi.org/10.1093/oso/9780198882121.001.0001

Shahbandeh, Mahsa (2024) 'Monthly Food Price Index Worldwide 2024',
    Statista. https://perma.cc/UU7P-EFT3

UN Trade & Development UNCTAD (n.d.) 'UN Trade & Development
    UNCTAD | UNCTADstat'. https://perma.cc/6WJS-BR3A

UN Trade & Development UNCTAD (2024) 'International Trade in Open
    and Transparent Markets May Help Alleviate the Effects of Shocks and
    Ensure Food Security, UN Trade & Development UNCTAD | SDG Pulse'.
    https://perma.cc/X38R-2V5H

WTO (World Trade Organization) (2023) 'Report on G20 Trade Measures
    (Mid-October 2022 to Mid-May 2023)', World Trade Organization.
    https://perma.cc/3CU5-H694

2. Africa's trade, food security and climate risks

Jamie MacLeod

This chapter aims to anchor the book in exactly what we mean when we consider Africa's agricultural trade – from grains and legumes through to fertilisers and tractors – and to establish a model for thinking about the interaction between trade, food security and climate risks in subsequent chapters.

It does this by examining Africa's agricultural exports in the broader context of its trade history. It then focuses on specific commodities such as maize, rice, wheat and fertilisers, which drive agricultural trade. The goal is to clarify the concept of 'food trade' as understood in this book.

The findings reveal that Africa's agricultural deficit has stabilised over the past decade in absolute terms despite rapid population growth. However, challenges in Africa's agricultural sector reflect broader issues seen in the continent's trade patterns. These include exporting unprocessed primary products in exchange for imports of finished consumer goods, thereby channelling Africa's raw materials towards value addition and processing jobs abroad rather than domestically. What is more, a worrisome dependence on imported food has emerged in several countries. This raises concerns about food security during shifts in trade terms and with the effects of climate change.

2.1 Five facts and a conceptual model of trade, food security and climate risks in Africa

The narrative surrounding Africa's food security, food trade and climate risks is intricate and defies reduction to a simple, all-encompassing story. Figure 2.1 illustrates the conceptual model used to frame the analysis of trade, food security and climate change in this book. Central to this framework are five narrative facts about Africa's agricultural trade.

How to cite this book chapter:

MacLeod, Jamie (2025) 'Africa's trade, food security and climate risks', in: Luke, David (ed) *How Africa Eats: Trade, Food Security and Climate Risks*, London: LSE Press, pp. 9–32. https://doi.org/10.31389/lsepress.hae.b License: CC-BY-NC 4.0

Figure 2.1: Conceptual model of trade, food security and climate change in Africa

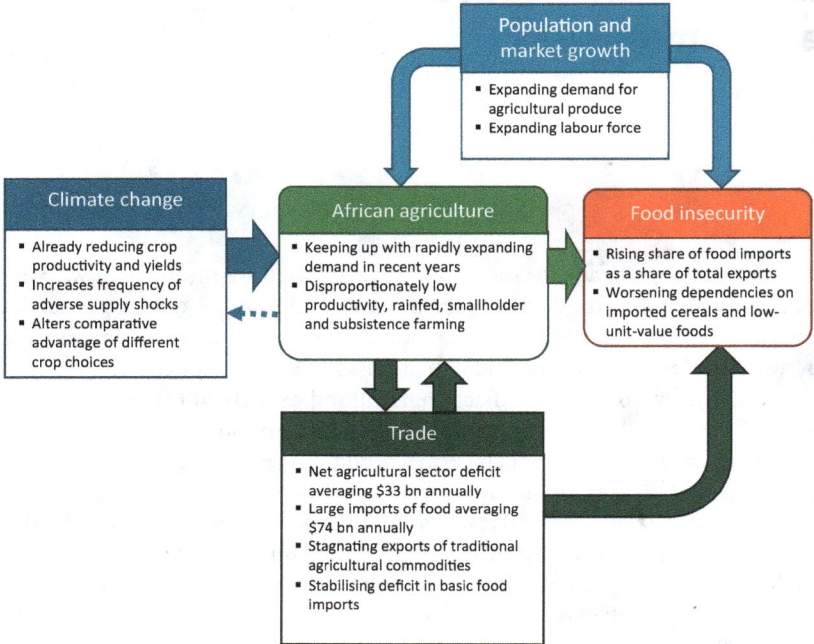

Source: Author's composition.

First, the continent faces a significant and persistent trade deficit when con-sidering the entire agricultural sector – including trade in foods, agricultural commodities, inputs and capital equipment. This deficit amounted to $33 billion in 2021, markedly higher than in the early 2000s, when the sector was at a breakeven and came close to even registering a small surplus.

Second, Africa is importing a tremendous amount of food, worth US$74 billion annually on average over the last five years. These food imports encompass low-unit-value items such as cereals and cooking oils, as well as higher-unit-value products such as fish and seafood, dairy and poultry. The demand for low-unit-value foods tends to increase with population growth, whereas higher-unit-value items correlate more closely with rising per capita incomes. In recent decades, population growth (a major confounding variable in our model) has outpaced other factors in putting increased pressure on food imports (Rakotoarisoa, Iafrate and Paschali 2011).

Third, exports of traditional African agricultural commodities have stagnated. Exports of agricultural commodities like cocoa, sugar, cotton, coffee, tobacco and tea have historically helped to balance out some of Africa's overall food trade deficit. In the post-independence period of the 1960s, these agricultural goods accounted for as much as 42 per cent of Africa's total merchandise exports (Saner, Tsai and Yiu 2012). However, such exports have since lagged, failing to keep pace with the continent's increasing food import bill, as well as growing more slowly than exports of manufactures or primary commodities.

Fourth, Africa's food import deficit has actually been stabilising. The deficit in Africa's agriculture sector has remained relatively stable over the past decade, decreasing from its peak of $47 billion in 2011, despite rapid population growth and rising per capita incomes during this period (Fox and Jayne 2020). This stability is encouraging news and means that Africa's agricultural sector has managed to keep pace reasonably well with rapidly expanding domestic demand, reversing a long-standing negative trend (Rakotoarisoa, Iafrate and Paschali 2011). Put another way, the uncontrolled, rising food import deficit experienced in the first decade of the millennium has been arrested.

Fifth, Africa's agricultural sector, population growth and associated trade patterns have important consequences for food security. The continent's approach to agricultural trade has led to an increasing reliance on food imports in several countries. Since the mid-2000s, North, East and West Africa have seen a rise in the share of total merchandise exports needed to fund their food import bills, positioning many African nations among the most dependent on food imports globally. On average, the median African country allocates a quarter of its export revenue to food imports, with 16 African countries spending over 40 per cent on food imports. These countries face substantial food insecurity in the event of adverse terms of trade shocks or global food price rises.

Of particular concern are staple foods like cereals, which are crucial for many regions. Around 30 per cent of the available cereal supply in African countries is imported, with North Africa experiencing the most severe dependency, importing 52 per cent of its cereals in 2017–2019 – a 10-percentage-point increase from 2003 to 2005.

What we can think of as these five 'narrative facts' describe an agricultural sector under stress – a sector in a stable trade deficit, driven by food imports, stagnating traditional commodity exports, and emerging risks, including food security.

Adding to these challenges is climate change, a formidable factor exacerbating existing agricultural and food security issues despite Africa's minimal contribution to global greenhouse gas emissions. Climate change is altering temperature averages and weather patterns, impacting optimal crop choices and increasing the frequency of supply disruptions. Yield projections indicate declines for staple crops across much of Africa, including vital sources of food security such as wheat, maize, sorghum and rice.

This cumulative effect underscores a pressing challenge that demands bold, ambitious and deliberate policy actions in agriculture and trade. The remainder of this chapter delves into the intricate details of this conceptual model – exploring specific products, countries and trade dynamics that shape Africa's trade, food security and climate resilience framework.

2.2 Africa's agricultural trade in perspective

Zooming out of just agriculture, the main story of Africa's overall trade is a persistent concentration in exports of primary commodities, particularly petroleum fuels but also metals, precious metals, and minerals. At the macroeconomic level, Africa's trade dynamics involves surpluses in exports of primary commodities that help to offset deficits in manufactures and the agriculture sector. Figure 2.2 illustrates this interplay, showing how expansions in primary commodity exports in some years mirror increases in imports of manufactured goods and basic food.

Figure 2.2: African countries' net exports by product categories, showing primary commodities exported for imports of manufactured and agricultural goods (US$ billion, current prices), 2002–2021

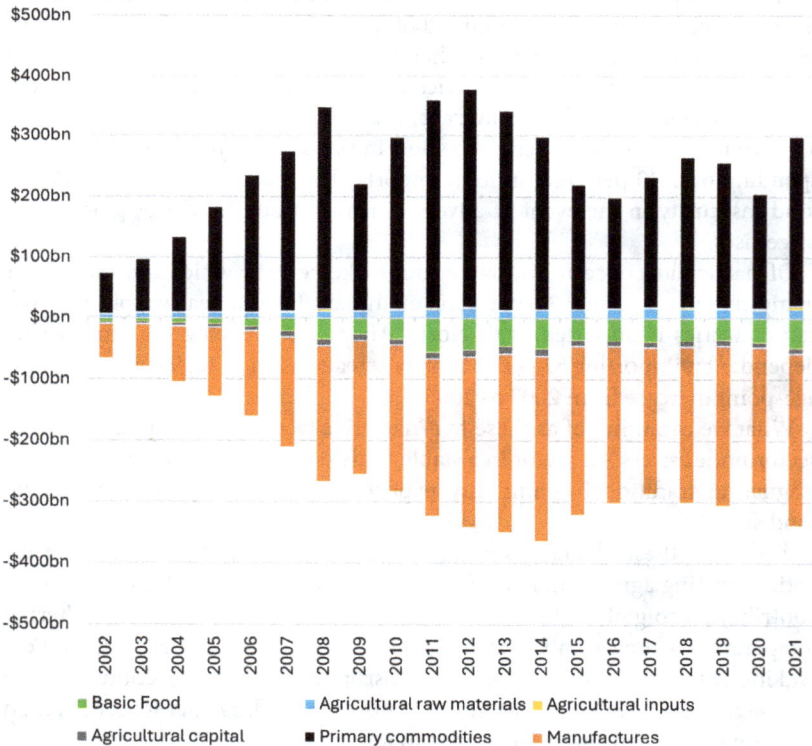

Source: Author's calculations based on Centre d'études prospectives et d'informations internationales's (CEPII) Base pour l'Analyse du Commerce International (Database for the analysis of international trade) (BACI) database (CEPII 2023; Gaulier and Zignago 2010). Notes: The continent collectively exports primary commodities in exchange for imports of manufactured goods and agricultural goods. See Annex A for details on product category composition. Negative values for net exports shown here refer to net imports.

As this book aims to understand Africa's agricultural trade holistically, it considers not just agricultural goods but also inputs such as fertilisers, as well as capital goods like tractors and agricultural equipment. Agricultural inputs are mostly fertilisers and herbicides. Agricultural capital goods include tractors, agricultural machinery – such as seeders, harvesters and dryers – and agricultural tools.

Africa's agricultural output is furthermore divided into two: basic foods and agricultural raw materials, as per the United Nations (UN) Conference on Trade and Development (UNCTAD) broad product group classifications. UNCTAD defines basic foods as edible products like grains, tubers, meat, fish, poultry, fruits and oil seeds. A classification and breakdown of these food products is provided in both the UN's Standard International Trade Classification (SITC) codes and in the World Customs Organization's Harmonized System, Chapters 1–24. However, the UNCTAD definition of basic foods excludes tropical beverages such as coffee and tea, and products such as cocoa, spices, vegetables, tobacco and alcoholic drinks that also fall within the SITC codes and the applicable chapters of the Harmonized System, several of which are among Africa's most important agriculture exports (UN trade & development 2024). On the basis of UNCTAD's definition of basic foods, cassava, yams, rice, maize, wheat, poultry, meat and fish, are identified in this book as the main products in Africa's basket of basic foods that are most widely consumed and thereby contribute most to daily calorie intake requirements. Meanwhile, agricultural raw materials encompass products like cotton, fresh cut flowers, wood, rubber, tea, coffee, cocoa, tobacco, and spices or flavourings such as vanilla. As we will see, the nature of this pre-established classification system can have some issues. Goods that in the African context behave more as commodities, such as raw cashew nuts, are counted as food exports.

While overshadowed by Africa's substantial deficit in manufactured goods, the overall import bill for the agricultural sector remains considerable, totalling $34 billion in 2021. Particularly stark is the deficit in 'basic food', which escalated to $49 billion in the same year. In contrast, Africa maintains a net surplus in agricultural raw materials, although this sector has not grown as well as other export segments over the past decade. Additionally, the continent relies on net imports of agricultural capital, such as machinery and tractors, while being a net exporter of agricultural inputs, notably fertilisers sourced predominantly from North Africa. South of the Sahara, however, Africa remains a net importer of agricultural inputs like fertilisers (author's analysis of United Nations n.d.).

Overall, the collective picture is of the agricultural sector exporting raw materials and commodities in exchange for imports of consumable foods, manufactures and capital goods. Figure 2.3 provides a detailed view of Africa's net trade dynamics within the agricultural sector. In 2021, African countries recorded a net trade deficit of $49 billion for basic foods and $9 billion for agricultural capital, while achieving net surpluses of $16 billion from exports of agricultural raw materials (including cocoa, tobacco, coffee, tea and spices)

Figure 2.3: African countries' net exports by product category in the agricultural sector (US$ billion, current prices), 2002–2021

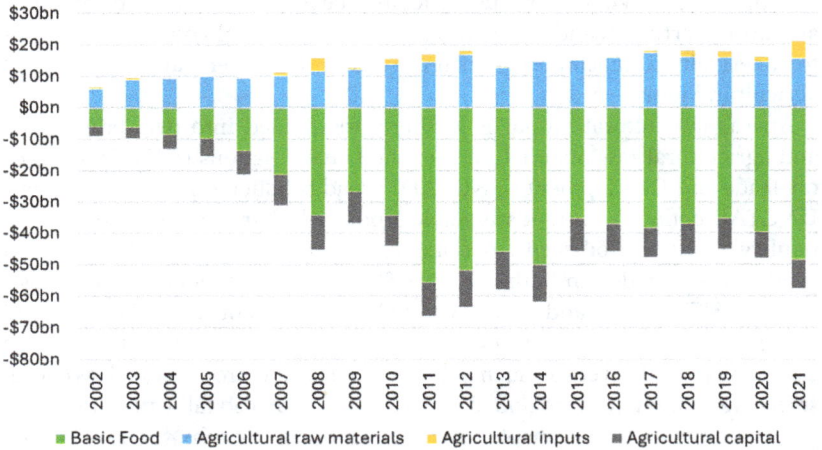

Source: Author's calculations based on CEPII's BACI database (CEPII 2023; Gaulier and Zignago 2010).
Notes: See Annex A for details on product category composition. Negative values for net exports shown here refer to net imports.

and of $6 billion from exports of agricultural inputs. This deficit widened dramatically from the early 2000s until 2011, before stabilising in the last decade, with the overall sector deficit in 2021 approximately 25 per cent smaller than in 2011. These figures are nominal and not adjusted for inflation, which means they underestimate the magnitude of the deficit observed in 2011.

Exchanging primary exports for processed imports in agricultural trade

Another telling way of making the same point is to redivide Africa's agricultural trade into that which is in its raw and unprocessed form, as opposed to processed goods that have been further worked and to which value has been added. This can be seen in Figure 2.4, which reconstitutes the individual products of Africa's trade using the UN's broad economic categories (BECs) to identify those that are primary or processed. This reformulation, for instance, casts cocoa beans, coffee beans, wheat and oranges as 'primary' goods, while classifying cocoa paste, roasted coffee, wheat flour and orange juice as 'processed'. A far larger share (56 per cent) of exports of agricultural produce from African countries are in a primary form, while 70 per cent of imports are processed.

Africa's agricultural trade is, in many ways, a microcosm of the challenges that Africa faces in its trade more broadly. African countries' exports are disproportionately concentrated in raw materials and unprocessed agricultural

Figure 2.4: Africa's exports and imports of food and agricultural raw materials, by BEC categories of primary and processed goods

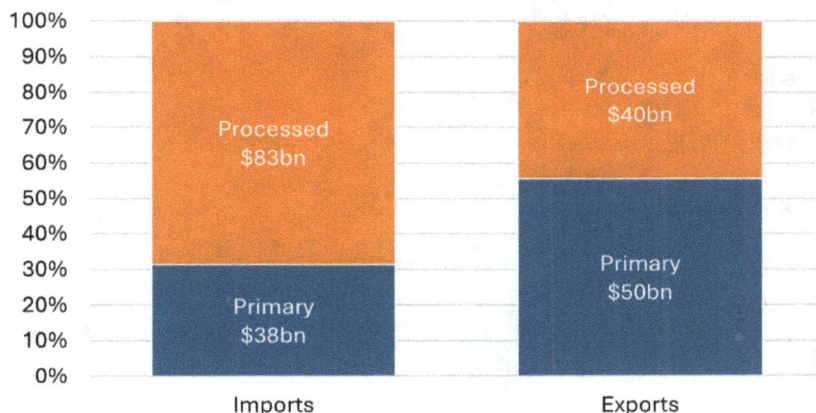

Sources: Author's calculations based on CEPII's BACI database (2023), based on a five-year average from 2017 to 2021.

exports. These often represent less value added than processed agricultural goods that have been further worked.

Who drives the agricultural deficit?

The agriculture deficit is of course not uniform across the continent. A large number of African countries have succeeded in emerging as important net exporters of foods and agricultural raw materials, chiefly Côte d'Ivoire, South Africa, Morocco, Ghana, Kenya and Uganda, while others lead very large deficits. This is shown in Figure 2.5, which plots each African country by the size of its net surplus (or deficit) in exports of food and agricultural raw materials along the y-axis. The country population, which is an important amplifying variable for the surplus (or deficit), is shown on the x-axis. The net agricultural exporters are in the top half of the chart.

The African countries in the bottom half of the chart are the culprits of agricultural deficits. The largest agricultural deficits tend to belong to countries with a combination of three things: large populations, like Egypt and Nigeria; a heavy focus on exports of primary commodities, such as Nigeria, Algeria, Angola and Libya; and those that are fragile and undermined by conflicts, like Somalia, Sudan and Democratic Republic of the Congo. The latter two factors (correspondingly shaded orange and red, respectively, in Figure 2.5) are suggestive of what matters for agricultural success: stability and an economy that is not too distracted by the trappings (or curses) of primary resources (Dauvin and Guerreiro 2017).

Figure 2.5: Net exports of food and agricultural raw materials, by country (US$ billion, current prices) and population

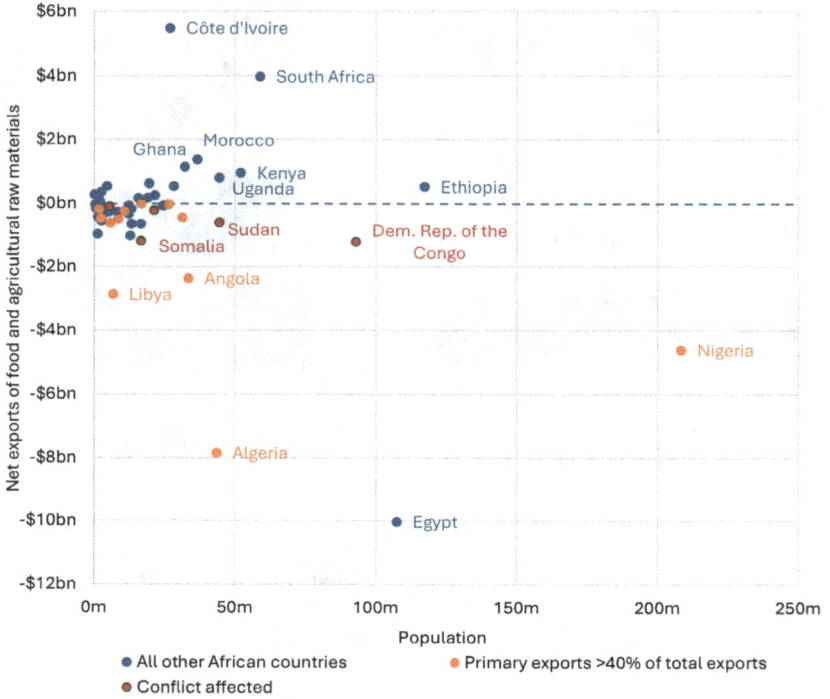

Sources: Author's calculations based on United Nations Department of Economic and Social Affairs (UN-DESA) (2022), CEPII's BACI database (CEPII 2023; Gaulier and Zignago 2010) and the World Bank (n.d) Classification of Fragile and Conflict-Affected Situations. Note: For simplicity, where a country is conflict-affected and at the same time its primary exports account for more than 40 per cent of total exports, it is classified as a primary exporter.

What exactly are Africa's agriculture sector exports?

Agricultural exports from African countries comprise an interesting blend of unprocessed, and lower-unit-value, agricultural commodities, like cocoa, sugar, cotton and coffee, but also higher-unit-value food products like fruits, nuts and fish.

What exactly are Africa's exports in each of these categories? Figure 2.6 breaks them down, showing the main products driving Africa's agricultural exports. Africa's major food exports, between 2017 and 2021, were fish and seafood ($9 billion); fruits ($7 billion) and citrus fruits ($3 billion); vegetables, roots and tubers ($6 billion); cashew nuts ($3 billion); sesame seeds ($2 billion); palm oil ($0.8 billion); and olive oil ($0.8 billion). Africa's exports of its major food security crops are worth less, with annual exports of maize amounting to $0.8 billion, wheat $0.5 billion and rice $0.4 billion.

Figure 2.6: Decomposition of Africa's average annual agricultural exports, by product category and product, five-year average (2017–2021)

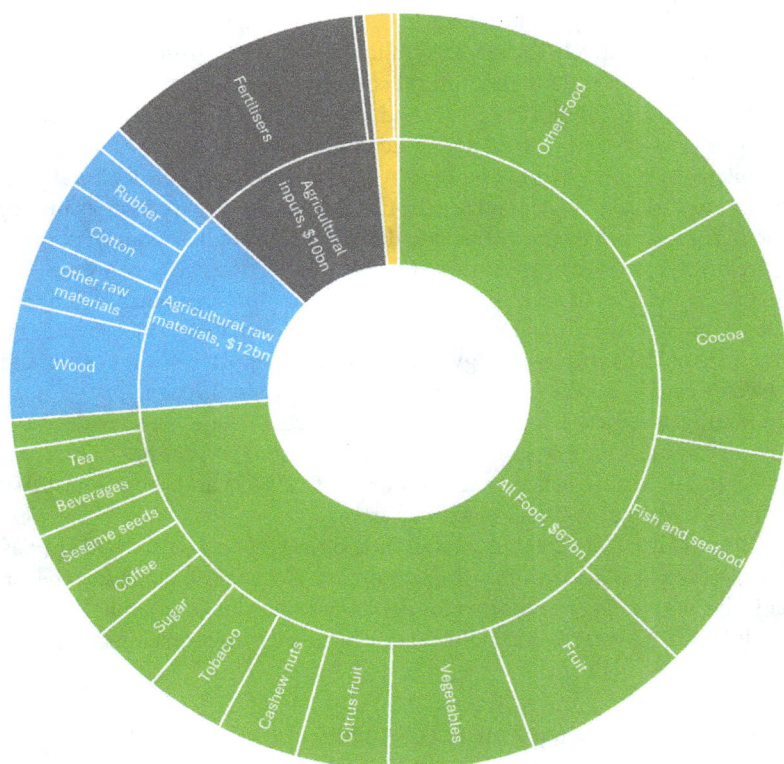

Source: Author's calculations based on CEPII's BACI database (CEPII 2023; Gaulier and Zignago 2010).
Note: See Annex for details on product category composition.

Africa's exports of agricultural commodities are dominated by cocoa ($10 billion), but sugar ($3 billion), cotton ($2 billion), coffee ($2 billion), tobacco ($2 billion), tea ($2 billion), flowers ($1 billion) and vanilla ($0.8 billion) are also important. Fertilisers are a sizeable agricultural input export ($10 billion), while total exports of agricultural capital amounted to just $0.6 billion. This unveils a story of African countries engaging far more in the upstream part of agricultural value chains and less so in the production or preparation of foods and goods for final consumption. This is the case even *within* broader categories of products like 'cashew nuts' or 'cocoa', with Africa's exports being more concentrated in raw cashew nuts than edible cashew kernels, and in cocoa beans and sugar more than processed cocoa or chocolate.

Who are Africa's leading food and agricultural commodity exporters and where do they export to?

Different African countries, and different export destinations, drive different parts of Africa's agricultural trade. The north and south of the continent tend to lead in relatively higher-unit-value food exports, while West and East Africa are more important agricultural commodity exporters.

Figures 2.7 and 2.8 illustrate the lead exporters, and destination markets, driving Africa's exports in food and agricultural commodities. This trade is led particularly by Northern and Southern Africa, by the relative agricultural powerhouses of Morocco, Egypt and South Africa, as well as a few countries in West Africa. This is shown by the thicker and darker lines from these countries, showing the relative value of bilateral export flows of foods from these countries.

In North Africa, Morocco is a major exporter of vegetables, fruits and seafood, mostly to Europe but also a few other destinations, like Russia, the United States (US) and Japan. Egypt is also a large exporter of citrus fruit and, to a lesser extent, vegetables to Europe, Russia and Saudi Arabia.

In Southern Africa, South Africa's food exports include citrus fruits (such as oranges and lemons) and other fruits (including avocados, berries and grapes) to Europe, as well as macadamia nuts to the US and several East Asia countries, as well as regional maize exports.

In West Africa, Côte d'Ivoire, Ghana and Nigeria (as well as Tanzania in East Africa) are major exporters of raw cashew nuts to Viet Nam and to India. The former is the world's biggest processor of raw cashew nuts into consumable cashew kernels, while the latter is one of the world's biggest cashew consumer markets.

Figure 2.7: Map showing Africa's top 50 bilateral exports of foods, between country pairs, five-year average (2017–2021)

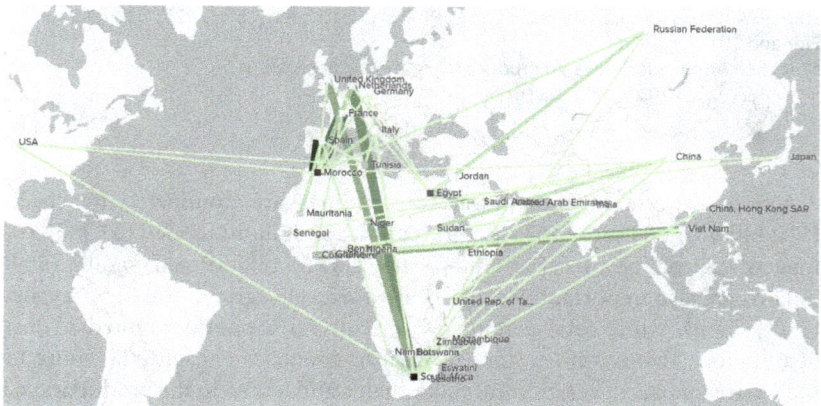

Source: Author's calculations based on CEPII's BACI database (CEPII 2023; Gaulier and Zignago 2010).
Note: Thicker and darker lines indicate the relative value of bilateral export flows of foods.

Other notable African food exports include olive oil from Tunisia, particularly to Spain and Italy, fish from Namibia and seafood from Mauritania to Spain. Also notable is China, which is a very important destination for sesame seeds from Sudan, Niger, Ethiopia, Togo and Tanzania, and groundnuts from Senegal.

Africa's exports of agricultural commodities are led much more by West Africa and Eastern Africa. This trade is dominated by exports of cocoa beans, from Côte d'Ivoire and Ghana and to a lesser extent Nigeria and Cameroon. This is mostly destined to Europe, but also to the US, Canada, Indonesia and Malaysia. After cocoa, cotton is the next most important agricultural commodity from West Africa, with Burkina Faso, Benin and Mali being notable exporters, especially to textile factories in Europe and Bangladesh.

In Eastern Africa, Kenya is a major exporter of tea to tea drinking markets such as Pakistan, Egypt, the United Arab Emirates and of course the United Kingdom. Kenya is also a major exporter of fresh cut flowers, particularly to the Netherlands, from where they are redistributed within Europe. Ethiopia and Uganda are large coffee exporters, mostly to Europe.

In Southern Africa, Malawi and Zimbabwe are notable tobacco exporters. Madagascar is a significant vanilla exporter, to the US and Europe. South Africa is the destination for regional exports of sugar from Eswatini, tobacco from Zimbabwe, and beef from Namibia and Botswana. South Africa itself also exports a large amount of sugar.

Beyond exports of food and agricultural commodities, African countries also export agricultural inputs and, to a much lesser extent, agricultural capital goods. In terms of agricultural inputs, North Africa (and especially Morocco) is a substantial exporter of fertiliser to countries all over the world.

Figure 2.8: Map showing Africa's top 50 bilateral exports of agricultural commodities, between country pairs, five-year average (2017–2021)

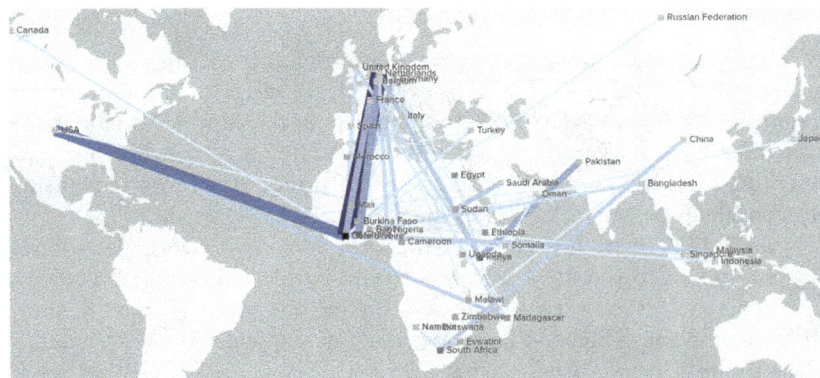

Source: Author's calculations based on CEPII's BACI database (CEPII 2023; Gaulier and Zignago 2010).
Note: Thicker and darker lines indicate the relative value of bilateral export flows of foods.

Table 2.1: Destination of Africa's agricultural exports, five-year average (2017–2021) ($bn)

	Europe	Intra-Africa	Asia	Americas	Other
Food	16	10	9	2	4
Agricultural commodities	12	5	6	3	3
Agricultural inputs	2	2	2	3	1
Agricultural capital	0.1	1	0.1	0.1	0.1
Total	**30**	**18**	**17**	**8**	**8**

Source: Author's calculations based on CEPII's BACI database (CEPII 2023; Gaulier and Zignago 2010).

With agricultural capital, South Africa is the main exporter within the continent, particularly of machinery and tractors to countries in Southern and East Africa. There is very limited regional trade in capital goods beyond those from South Africa.

Table 2.1 shows how these trade flows aggregate into Africa's exports of agricultural goods to different markets. The European Union (EU) remains by far Africa's most important market for exports of food and agricultural commodities. Intra-African trade is also very important, and dominated by trade in food products, while involving less trade in agricultural commodities. This trade also includes a large informal component that is not reflected in the data presented here (informal intra-African food trade is discussed in Chapter 4). In contrast, agricultural commodities comprise a large share of Africa's agricultural exports to countries in Asia, the Americas and elsewhere in the world.

What are Africa's agricultural imports?

Africa's agricultural imports are far more concentrated in foods. Food accounts for two-thirds of all African agricultural imports, with Africa importing far fewer agricultural commodities than it exports. Even Africa's agricultural commodity imports are dominated by sugar, which, while representing a commodity that will be processed further when exported outside the continent, is often a direct consumer good when imported.

From 2017 to 2021, African countries averaged annual food imports totalling $73 billion. This trade is primarily fuelled by substantial imports of low-cost items such as food security cereals, cooking oils and sugar. Additionally, there are imports of higher-value goods like fish, seafood and dairy products. These imports include products that Africa produces competitively but not at a scale sufficient to satisfy its substantial and expanding consumer demand, necessitating significant importation. Figure 2.9 provides a detailed breakdown of Africa's agricultural import patterns.

Figure 2.9: Decomposition of Africa's average annual agricultural imports, by product category and product, five-year average (2017–2021)

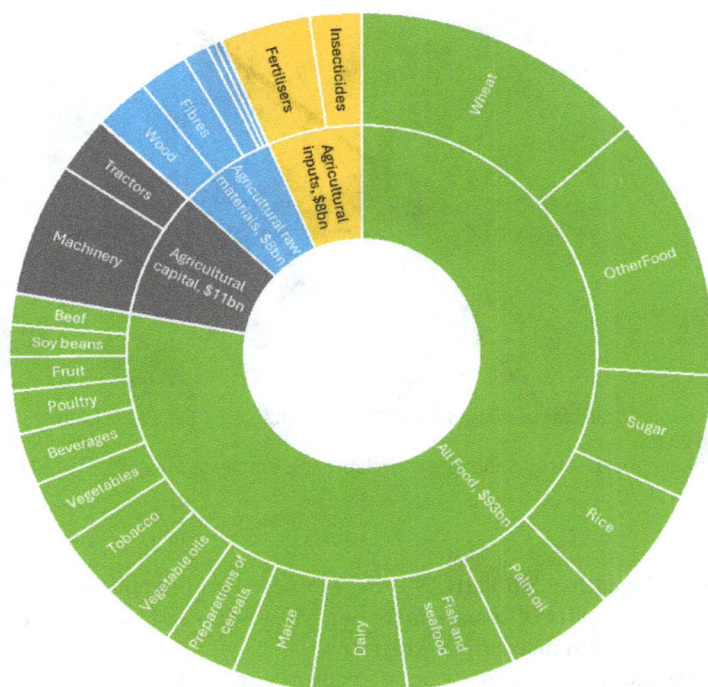

Source: Author's calculations based on CEPII's BACI database (CEPII 2023; Gaulier and Zignago 2010).
Note: See Annex for details on product category composition.

Where do the food imports come from and to which African countries do they go?

Figure 2.10 provides a visual representation of Africa's significant food trade flows, detailing the top 50 bilateral import relationships. It highlights that Africa sources its food from a diverse array of countries across all continents, contrasting with its export focus on foods and agricultural commodities primarily with Europe. While regional trade, particularly in Southern Africa, involves grains like maize exported from South Africa to neighbouring countries, the majority of Africa's major food import partners are located outside the continent.

North Africa plays a pivotal role in Africa's food imports, with Egypt notably importing wheat from Russia and Ukraine, maize from Brazil and Argentina, soya beans from Argentina and the EU, palm oil from Indonesia, and

Figure 2.10: Map showing Africa's top 50 bilateral imports of foods between country pairs, five-year average (2017–2021)

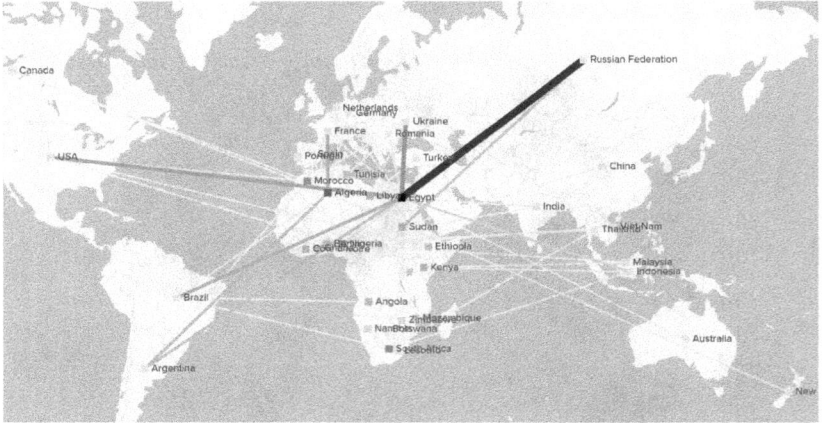

Source: Author's calculations based on CEPII's BACI database (CEPII 2023; Gaulier and Zignago 2010).
Note: Thicker and darker lines indicate the relative value of bilateral export flows of foods.

beef from Argentina. Similarly, Algeria, Morocco, Tunisia and Sudan import substantial quantities of cereals, including wheat sourced from Canada, Russia and the EU, and maize from Argentina.

In West and Central Africa, rice and palm oil are the predominant imports. Rice originates from countries such as India, Viet Nam and Thailand, while palm oil is sourced from Malaysia and Indonesia. Nigeria stands out as the largest importer of food in West Africa, driven by its sizeable population, importing significant quantities of wheat from the US, Canada and Russia, as well as rice and palm oil from Asia. South Africa similarly imports substantial amounts of rice from Thailand and India, palm oil from Indonesia and Malaysia, and beef from Brazil.

East Africa, while representing a smaller share of major food inflows into the continent, sees notable imports of palm oil, particularly by Kenya and to a lesser extent Ethiopia, from Indonesia and Malaysia.

Figure 2.10 does not depict Africa's imports of agricultural commodities, which are dominated by sugar imports, notably from Brazil (the world's largest sugar exporter) and to a lesser extent India. Other noteworthy agricultural commodities include the regional trade in tea, such as from Kenya to Egypt, and the tobacco trade from Zimbabwe to South Africa.

Imports of agricultural capital goods are also important. South Africa is a major destination for tractors imported from Germany, Sweden and Brazil. In West Africa, tractors are predominantly sourced from China, while North African countries rely on European suppliers for their tractor imports.

Table 2.2: Origin of Africa's agricultural imports, five-year average (2017–2021) ($bn)

	Europe	Asia	Americas	Intra-Africa	Other
Food	27	17	13	10	5
Agricultural commodities	4	3	6	5	1
Agricultural inputs	2	2	0	2	1
Agricultural capital	5	3	1.1	1	1
Total	**38**	**25**	**20**	**18**	**8**

Source: Author's calculations based on CEPII's BACI database (CEPII 2023; Gaulier and Zignago 2010).

A striking observation from the data is the substantial role of China as a supplier of agricultural capital goods, including tractors and machinery, to various countries across Africa. Additionally, there is notable regional trade dynamics, with South Africa serving as a supplier of agricultural capital goods to its neighbouring countries.

The most significant flows of these are tractors into South Africa from Germany, Sweden and Brazil; tractors into West African countries from China; and tractors into North African countries from Europe. Many African countries also import fertilisers including large quantities of fertilisers sourced primarily from Morocco, the Middle East and Russia, as well as insecticides imported from China.

Table 2.2 summarises the origin of Africa's imports of agricultural goods. Europe is by a large margin the most important source of these, followed by Asia, the Americas and then intra-African trade. However, African countries import mostly food products from Europe and Asia, and especially lower-unit-value food goods like grains and palm oil.

Agricultural imports and Africa's food security

As demonstrated, Africa faces a significant challenge with its substantial food imports, resulting in a large annual net deficit. Often, these imports are financed using foreign exchange earned from other export-competitive sectors of African economies, such as primary commodities. However, dependence on imported foods introduces risks, particularly when these imports constitute a large portion of foreign exchange earnings. Changes in trade terms, such as declines in export prices or increases in global food prices, can jeopardise food security. This risk is amplified when there is a deficit in the trade of basic foods, especially in terms of calories rather than financial value, making it impractical to redirect exported foods to meet domestic consumption needs. In 2023, at least 34 African countries were grappling with such challenges, according to the FAO (n.d.).

Figure 2.11: Value (in percentage) of food imports in total merchandise exports (three-year average)

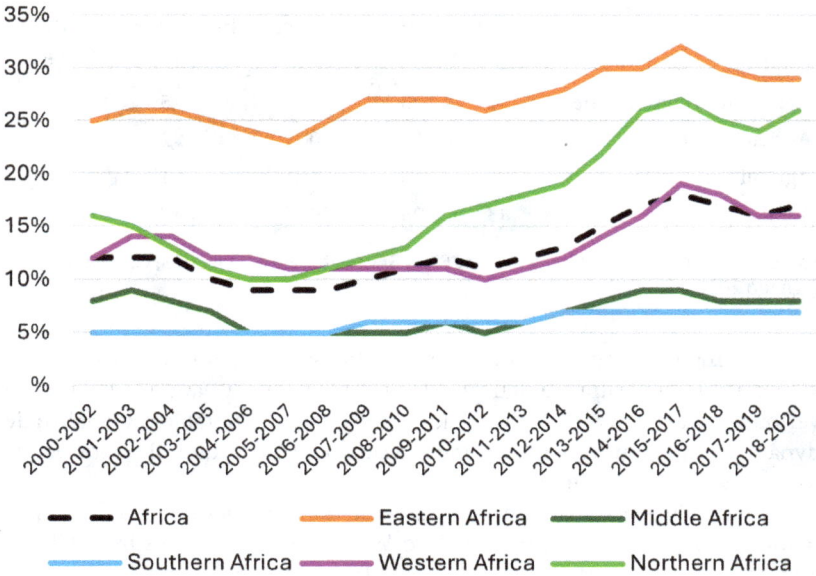

Source: Calculated on the basis of FAO (2023).
Notes: The indicator is calculated in three-year averages, from 2000–2002 to 2018–2020, to reduce the impact of possible errors in estimated trade flows. The aggregates are computed by weighted mean, using total merchandise trade as weighting variable. Value of food (excluding fish).

Figure 2.11 illustrates the value of food imports as a percentage of total merchandise exports for Africa as a whole and its five regions. This metric serves as a gauge of vulnerability, reflecting the adequacy of foreign exchange reserves to cover food imports and its implications for national food security based on production and trade patterns.

Africa's reliance on food imports as a percentage of total merchandise exports has steadily risen since 2003, reaching 17 per cent during the 2018–2020 period. In comparison, the global average for this indicator was 7 per cent during the same period. East Africa emerges as the most vulnerable region, with 29 per cent of export earnings allocated to food imports in recent years. North Africa has experienced the most pronounced deterioration, increasing from 10 per cent in the 2002–2004 period to 26 per cent in 2018–2020. West Africa has also seen a notable increase in the value of food imports as a proportion of total exports. Cumulatively, these trends underscore the heightened susceptibility of Africa's food security to shocks in terms of trade.

As depicted in Figure 2.12, numerous African countries face severe food import insecurity, as evidenced by the high proportion of food imports relative to total merchandise exports. This vulnerability is particularly acute in

Figure 2.12: Map showing worldwide distribution of food import insecurity as reflected in percentage of food imports in total merchandise exports by value, 2018–2020

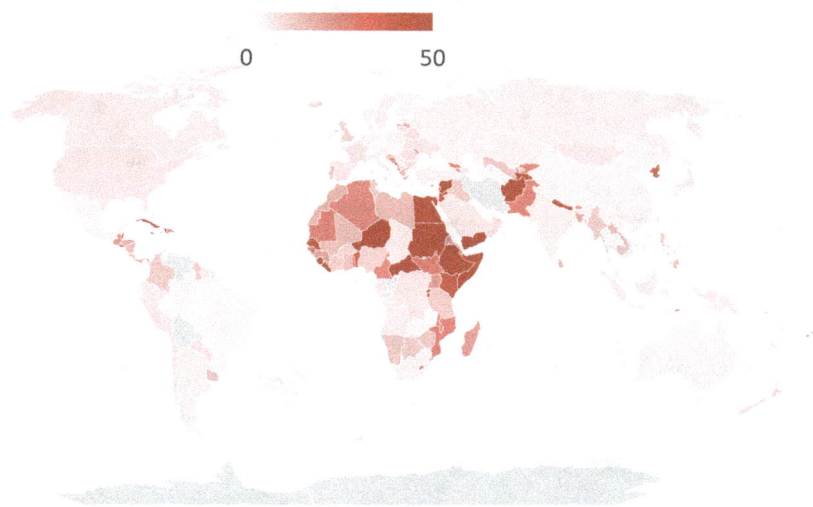

0   50

Powered by Bing
© Australian Bureau of Statistics, GeoNames, Microsoft, Navinfo, OpenStreetMap, TomTom, Zenrin

Source: Calculated on the basis of FAO (2023).
Note: The indicator is capped at >50 to limit the visual impact of outliers.

countries across the Horn of Africa (Somalia, Ethiopia and Sudan), North Africa (Egypt) and several smaller nations in West, East and Central Africa (including Sierra Leone, Liberia, Benin, Gambia, Cabo Verde, Sao Tome and Principe, Niger, Senegal, Comoros, Guinea-Bissau, Burundi and Central African Republic). In these countries, more than 40 per cent of their export earnings are allocated to financing food imports. Cereals emerge as the critical crop for food security across most of these nations.

Figure 2.13 illustrates Africa's cereals import dependency ratio over time, offering insights into the extent to which countries rely on imported cereals compared to their domestic production. This ratio is calculated as (cereals imports − cereals exports)/(cereals production + cereals imports − cereals exports) * 100. The indicator's values are capped at 100, indicating complete dependence on imports when exceeded.

These metrics underscore the significant challenges many African countries face in ensuring food security, with a heavy reliance on imported cereals in nations already struggling with high proportions of export earnings allocated to food imports.

Around 30 per cent of the available food supply of cereals in African countries is sourced through imports. This is in stark contrast to the world average,

Figure 2.13: Cereals import dependency, across Africa and over time: value (in percentage) of imported cereals in total available supply of food cereals (three-year average)

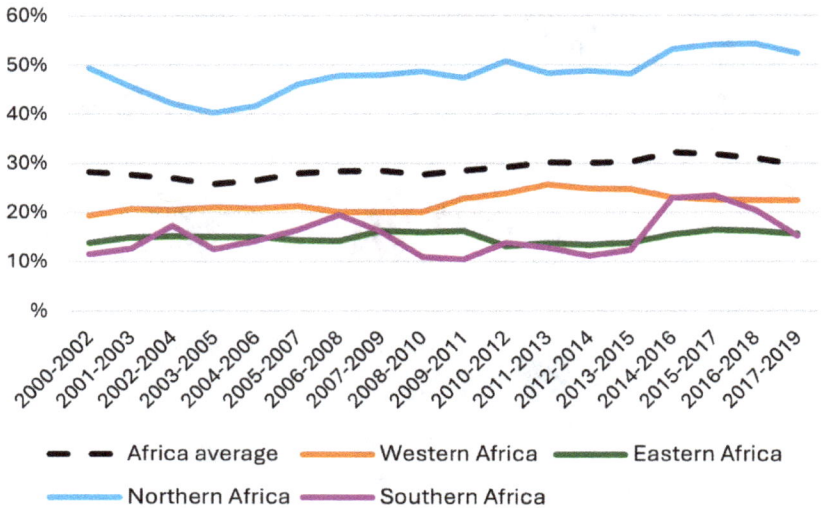

Source: Calculated on the basis of FAO (2023).
Note: The indicator is calculated in three-year averages, from 2000–2002 to 2017–2019, to reduce the impact of possible errors in estimated production and trade, due to the difficulties in properly accounting of stock variations in major food.

which stands at just under zero (−2.8 per cent during 2017–2019, indicating that most countries are net exporters of cereals). The dependency on imported cereals is most pronounced in North Africa, where 52 per cent of the available cereals are imported. This reflects a deterioration of 10 percentage points compared to the period from 2003 to 2005. Figure 2.14 puts the cereals import dependency of African countries in the context of the world. As a continent, it is clear that Africa is the part of the world that struggles most with a dependency on imported cereals, alongside the Arabian Peninsula.

2.3 The impact of climate change on agriculture, food security and trade in Africa

We can think of climate change as interacting with trade and impacting food security and agriculture in Africa through two primary channels. This simplification helps us to think through the key role of trade, though in practice climate change impacts are multidimensional, intersecting with socio-economic, political and environmental factors such as security, migration and labour productivity.

Figure 2.14: Global cereals import dependency as reflected in value (in percentage) of imported cereals in total available supply of food cereals (2017–2019)

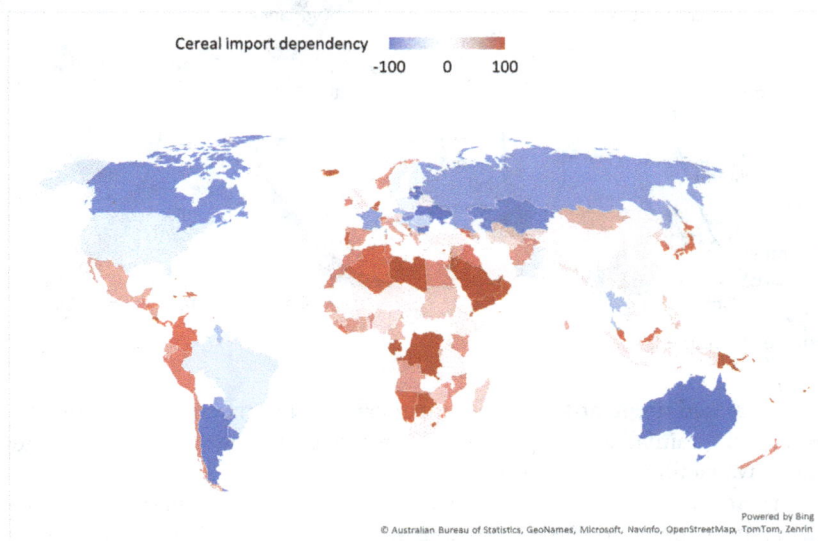

Source: Calculated on the basis of Food and Agriculture Organization of the United Nations (n.d.).
Notes: The indicator is capped at −100 so as not to distort the visualisation with outlier values. Negative values indicate that a country is a net exporter of cereals.

The first is by 'shifting the mean' in weather systems. Climate change is increasing temperatures and altering precipitation patterns. Rising temperatures can lead to changing growing seasons or alter the suitable geographical range for specific crops. In some regions, the temperature increase may more frequently exceed the optimal range for certain crops, negatively impacting yields. The first stylistic histogram of Figure 2.15 shows the frequency of cold and hot weather events under a normal weather distribution function. Climate change shifts the mean temperature, which in turn increases the frequency of hot weather events. In this way, the 'mean' weather system would be considered to have shifted, with consequences for yields of both agricultural commodities and foods. Such changes have already reduced maize and wheat yields in sub-Saharan Africa, with yields projected to fall for most staple crops across most of Africa, alongside declines in livestock production and marine and freshwater fisheries (Intergovernmental Panel on Climate Change (IPCC) 2023).

The second channel of impact from climate change to agriculture concerns 'shifting the variance'. This is demonstrated in the second stylistic histogram in Figure 2.15. A larger variance refers to increases in the occurrence of extreme weather events, such as extreme heat stress, floods or droughts.

Figure 2.15: Stylistic weather histogram

a) Weather volatility: shifting the mean   *b) Weather volatility: shifting the variance*

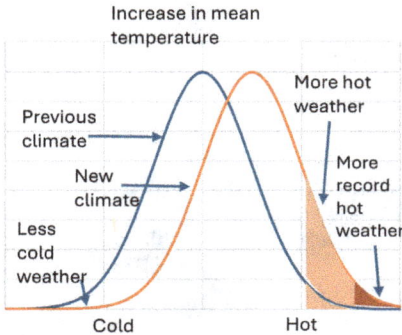

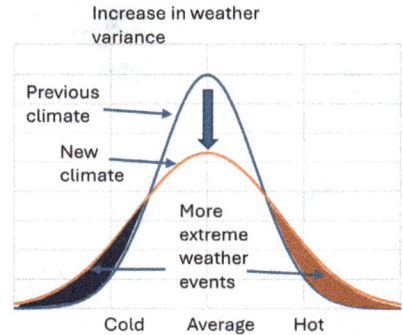

Source: Author's elaboration.

The increased irregularity of weather patterns, including prolonged droughts and intense rainfall events, poses challenges for crop growth, soil fertility and water availability. Another way we might think of this is as 'weather shocks'. Climate change is causing increases in drought frequency and duration over large parts of Southern Africa, flooding in North Africa, and heat waves across the continent (Intergovernmental Panel on Climate Change (IPCC) 2023).

Each channel of impact affects both the optimum choice of crop and the frequency of adverse supply shocks and, in turn, food security challenges. Trade interacts with these changes in two key ways: by allowing a safety valve for food availability and stability in the presence of adverse supply shocks and by allowing trade to evolve to reflect changing optimum growing parameters.

Gradual changes in climate will see a change in the agricultural comparative advantage of different countries. For instance, rising temperatures may cause certain crops to be less efficiently produced in one country but more efficiently produced in another, thereby changing the comparative advantage of each country with respect to one another. Driving these changes is the impact that climate change has on agricultural productivity. Maize and wheat yields have already decreased an average of 5.8 per cent and 2.3 per cent, respectively, between 1974 and 2008, in sub-Saharan Africa owing to climate change (Intergovernmental Panel on Climate Change (IPCC) 2023). Trade can become an important tool for food security in these instances by allowing consumers to adapt more readily to changing market conditions by tapping world markets while allowing producers to grow what reflects their changing comparative advantage (Baldos and Hertel 2015).

Trade can be used to mitigate the impact of production shocks, including those affecting critical food security crops. This is because adverse supply shocks in certain places can be met by supply surpluses in other, unaffected places, through trade. Staple food crops tend to have greater price volatility the more remote and detached from global markets they are (Burgess et al.

2011, p.26; Moctar et al. 2015). Markets that are better connected and more open can help mitigate the severity of supply shocks for agricultural products, including those that are climate-induced. However, availability is only part of any solution, with food security requiring other aspects of purchasing power to ensure meaningful access.

Beyond trade in actual agricultural produce, trade in agricultural intermediates and inputs, as well as agricultural services and knowledge, can play an important role in agricultural adaptation to climate change. Improving access can help farmers utilise new seed varieties, agricultural machinery, fertilisers and agricultural extension services to address changing climate challenges. African farmers pay considerably more for fertilisers than farmers in countries like Pakistan, Argentina and Brazil do (Keyser 2012). As well as reducing formal tariff barriers, alleviating non-tariff barriers would be important here. Ways to do this include regional harmonisation, or mutual recognition, of standards and seeds certification, improving competition between logistics suppliers, aiding the mobility of agricultural specialists, and reducing opaque and unpredictable trade policies. Policy measures that could be taken to discipline non-tariff barriers are discussed in Chapter 7.

Conclusions

The agricultural sector reflects broader challenges within Africa's trade dynamics, characterised by a pattern of exchanging raw, unprocessed exports for imports of final consumption goods. This scenario is fuelled by the primacy of primary commodities and the persistence of conflict and fragility on the continent.

This trade dynamic primarily benefits value addition and material processing industries outside Africa rather than domestically. The trend is starkly evident in Africa's exports of primary commodities such as fuels and metals in exchange for manufactured goods. Similarly, Africa's agricultural sector follows a parallel trade pattern: while final consumer goods, particularly food items, are imported, exports predominantly consist of intermediate goods such as cocoa beans, cotton, raw cashew nuts and fertilisers – essential inputs for production and agricultural processing elsewhere in the world.

This trade structure, akin to Africa's exports of petroleum oils and metals, does not promote local value addition, job creation or economic growth. Many Africans remain trapped in unproductive or informal employment, missing out on opportunities for higher earnings that could arise from processing Africa's agricultural and food commodities within the continent. Moreover, several of Africa's primary agricultural and food commodities are expected to see declining prices relative to manufactured goods (Harvey et al. 2010).

The current structure of Africa's agricultural trade puts the continent at risk to emerging and continuing threats. Too many African countries are dependent on imports of staple foods for sustenance. Many spend a large share of their export earnings on these imports. Supply or adverse terms of trade shocks could prove perilous for these countries. These challenges are amplified by the pressures of climate change. Frank questions and bold answers are needed to policies in both trade and agriculture. In order to understand the terrain of Africa's agricultural trade, the chapters of this book now turn to these policy consequences.

References

Baldos, Uris L. C.; and Hertel, Thomas W. (2015) 'The Role of International Trade in Managing Food Security Risks from Climate Change', *Food Security*, vol. 7, no. 2, pp.275–90. https://doi.org/10.1007/s12571-015-0435-z

Burgess, Robin; Deschenes, Olivier; Donaldson, Dave; and Greenstone, Michael (2011) *Weather and Death in India*. Cambridge, MA: Massachusetts Institute of Technology, Department of Economics.

Dauvin, Magali; and Guerreiro, David (2017) 'The Paradox of Plenty: A Meta-Analysis', *World Development*, vol. 94, pp.212–31. https://doi.org/10.1016/j.worlddev.2017.01.009

FAO (n.d.) 'Low-Income Food-Deficit Countries (LIFDCs) – List updated June 2023', Food and Agriculture Organization of the United Nations. https://perma.cc/VQ2V-7MCN

Food and Agriculture Organization of the United Nations (n.d.) 'FAOSTAT'. https://perma.cc/5Y7P-WM75

Fox, Louise; and Jayne, Thomas S. (2020) 'Unpacking the Misconceptions about Africa's Food Imports', Brookings. https://perma.cc/V69G-F6PP

Gaulier, Guillaume; and Zignago, Soledad (2010) 'BACI: International Trade Database at the Product-Level. The 1994-2007 Version', Working Papers 2010–23. CEPII. https://perma.cc/X42W-8NCX

Harvey, David I.; Kellard, Neil M.; Madsen, Jakob B. and Wohar, Mark E. (2010) 'The Prebisch-Singer Hypothesis: Four Centuries of Evidence', *The Review of Economics and Statistics*, vol. 92, no. 2, pp.367–77. https://doi.org/10.1162/rest.2010.12184

IPCC (Intergovernmental Panel on Climate Change) (ed.) (2023) 'Africa', in: *Climate Change 2022 – Impacts, Adaptation and Vulnerability: Working Group II Contribution to the Sixth Assessment Report of the Intergovernmental Panel on Climate Change*, Cambridge: Cambridge University Press, pp.1285–456. https://doi.org/10.1017/9781009325844.011

Keyser, John (2012) 'Africa Can feed Africa: Removing Barriers to Regional Trade in Food Staples', Background paper for Africa Can Help Feed

Africa: Removing Barriers to Regional Trade in Food Staples, edited by P. Benton, Washington, DC: World Bank.

Moctar, Ndiaye; Elodie, Maitre d'Hôtel; and Tristan, Le Cotty (2015) 'Maize Price Volatility: Does Market Remoteness Matter?', World Bank Policy Research Working Paper (7202).

Rakotoarisoa, Manitra A.; Iafrate, Massimo; and Paschali, Marianna (2011) *Why Has Africa Become a Net Food Importer?* Rome: Food and Agriculture Organization of the United Nations. https://perma.cc/A5VL-Q6MP

Saner, Raymond; Tsai, Charles; and Yiu, Lichia (2012) 'Food Security in Africa: Trade Theory, Modern Realities and Provocative Considerations for Policymakers', *Governance, Regional Integration, Economics, Agriculture and Trade*, September, pp.17–18. https://perma.cc/5XJB-ZGW6

UN trade & development, UNCTAD (2024) 'Merchandise Trade Matrix, annual' UNCTAD, Data Hub. https://unctadstat.unctad.org/datacentre/dataviewer/US.TradeMatrix

United Nations Department of Economic and Social Affairs (UN-DESA), Population Division (2022) *World Population Prospects: The 2022 Revision.*

World Bank (n.d.) *Classification of Fragile and Conflict-Affected Situations* World Bank. https://perma.cc/F783-D95K

World Bank Open Data (n.d.) 'World Bank Group | Data'. https://data.worldbank.org

Annex: categorisation of agricultural trade

Category	SITC codes (unless otherwise specified)
Manufactures	5 – Chemicals and related products, except 56 (crude fertilizers) 6 – Manufactured goods classified chiefly by material, except 68 (nonferrous metals) and 667 (pearls, precious and semi-precious stones, worked or unworked) 7 – Machinery and transport equipment 8 – Miscellaneous manufactured articles
Primary commodities	3 – Mineral fuels, lubricants and related materials 667 – Pearls, Precious And Semi-Prec. Stones 68 – Non-Ferrous Metals 971 – Gold, Non-Monetary
Agricultural raw materials	07 – Tobacco, cocoa, tea, coffee and spices 2 – Crude Materials, except 22 (Oil seeds and oleaginous fruits), 27 (fertilizers), and 28 (Metalliferous Ores And Metal Scrap)

(Continued)

Annex: (Continued)

Category	SITC codes (unless otherwise specified)
Basic food	0 – Food and live animals, except 07 (tobacco, cocoa, tea, coffee and spices) 1 – Beverages and tobacco 22 – Oil seeds and oleaginous fruits 4 – Animal and vegetable fats
Agricultural inputs	SITC codes 271/272 – Crude fertilisers 56 – Fertilisers Harmonised System (HS) codes 380810 – Insecticides 380820 – Fungicides 380830 – Herbicides 380890 – Rodenticides
Agricultural capital goods	HS codes 401161 – Tractor tyres 401192 – Tractor tyres 820140 – Hand tools for agriculture, horticulture or forestry 820190 – Hand tools for agriculture, horticulture or forestry (n.e.s.) 820840 – Knives and blades used for agriculture, horticulture or forestry 841931 – Dryers for agricultural products 842121 – Centrifuges for filtering or purifying liquids 842481 – Liquid spraying equipment for agricultural or horticultural use 842890 – Loaders (e.g. for agricultural tractors) 8432 – Agricultural machinery for soil preparation, seeders or cultivation 8433 – Harvesting or threshing machinery 8434 – Milking machines and dairy machinery 8435 – Presses, crushers and similar machinery used in the manufacture of wine, cider, fruit juices or similar beverages 8436 – Other agricultural, horticultural, forestry, machinery used for preparing animal feeding stuffs and poultry incubators 8437 – Machinery for cleaning, sorting or grading seed, grain or dried leguminous vegetables 8438 – Machinery for the industrial preparation or manufacture of food or drink 847920 – Machinery for the extraction or preparation of fats and oils 8701 – Agricultural tractors 871620 – Trailers and semi-trailers for loading agricultural produce 940600 – Greenhouses

Source: Author's calculations.

3. What Africa eats – the basic foods

Olawale Ogunkola and Vinaye Dey Ancharaz

This chapter examines food production and consumption in Africa in relation to the eight basic foods that we identify as essential for food security on the continent. Two are tubers (cassava and yams), three are cereals (maize, rice and wheat) and three are animal proteins (meat, poultry and fish). Following a comparative overview that situates Africa in relation to world food production, the core sections of the chapter examine temporal and spatial patterns in the production and consumption of the selected basic foods, drawing upon data (FAOSTAT) from the Food and Agriculture Organization of the United Nations (FAO). The chapter concludes with a discussion of ways in which climate change interacts with the basic foods and emerging adaptation and mitigation measures building upon the model of trade, food security and climate change identified in Chapter 2.

3.1 Africa in world food production and consumption

Africa, with 17 per cent of the world population, contributes about 8 per cent of global food production. In the 30 years between 1991 and 2021, the value of global food production increased at an average annual growth rate of 2.4 per cent, from $1.02 to $4.2 trillion. However, the world annual population growth rate was 0.80 per cent in 2022. Over the same period, Africa's food production grew at 2.1 per cent, below the continent's population growth rate of 2.4 per cent. This is not because Africa specialises in industries other than agriculture, since more of its economy is centred around agriculture than in the rest of the world (in 2022, agriculture, forestry and fishing accounted for 15 per cent of the continent's gross domestic product (GDP), compared to only 4 per cent for the world as a whole). It is because productivity in

---

Figure 3.1: Food production value (US$ billion, current prices), Africa and the world, 1991–2021

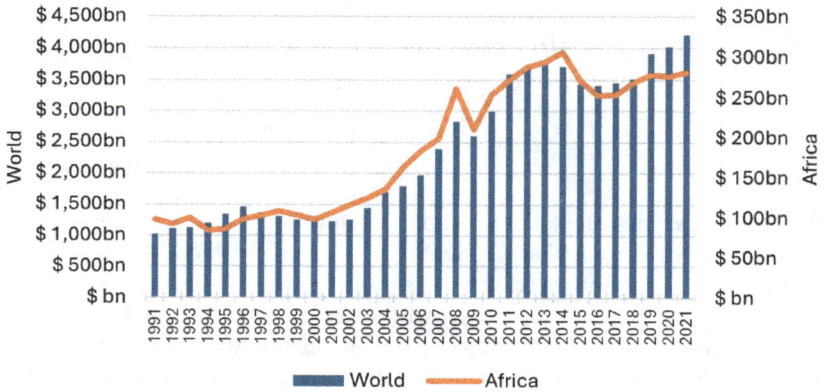

Source: Authors based on data from FAOSTAT.

Figure 3.2: Gross Food Production Value Index, Africa and the world, 1991–2021 (1991 = 100)

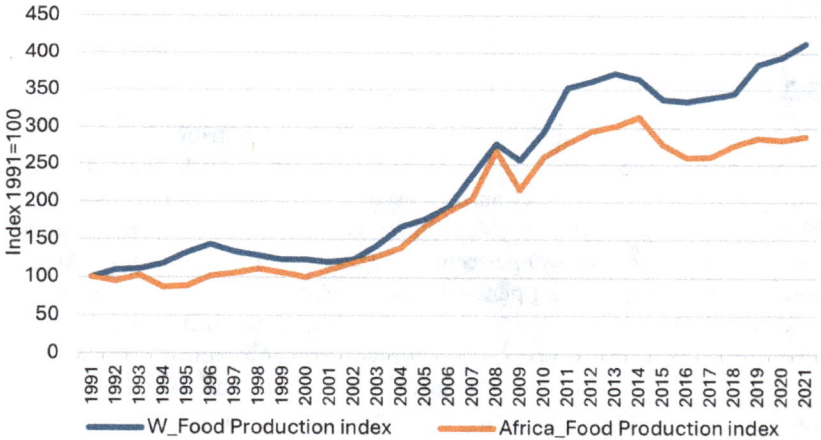

Source: Authors based on data from FAOSTAT.

agriculture (and the economy as a whole) is much lower than in the rest of the world (authors' analysis of World Bank Open Data n.d.).[1]

Figures 3.1 and 3.2 present these trends in global and Africa food production in value and index, respectively. Figure 3.1 shows world food production (left hand axis) quadrupling from $1 trillion in 1991 to $4.2 trillion in 2021 in current prices. The value of Africa's food production

Figure 3.3: Food production (US$ billion, current prices) by African subregion, 1991–2021

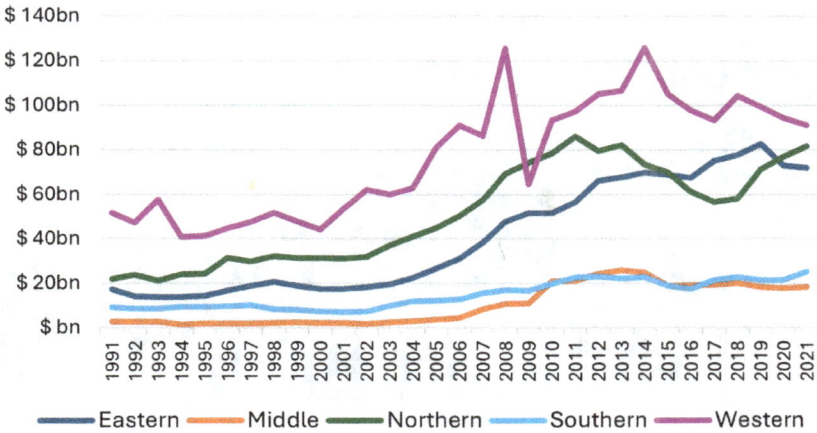

Source: Authors based on data from FAOSTAT.

(right axis of Figure 3.1) also increased from US$92.3 billion in 1991 to US$281.2 billion in 2021 in current prices, but below the global trend in the three decades.

Figure 3.2 presents global and Africa food production indexes, with the 1991 level set at 100. The indexes trended together until 2008, when the world food production index was 278.13 and the corresponding figure for Africa was 268.19. However, since 2008, a gap has emerged between the two and this was maintained even during the 2020–2021 Covid-19 shock. Although drawn from different data sets, these trends against the demographic background are consistent with headline data on food security that were introduced in Chapter 1.

Africa is of course far from being homogeneous. While regional variations may hide country differences, some broad contours are discernible from regional food production patterns. Figure 3.3 shows the trends in the value of Africa's food production by region. The shares of the different regions in food production by value are shown in Figure 3.4. All regions recorded an overall increase in the value of food production over the period, with population, policy, productivity, conflict, climate and ecological factors accounting for some of the fluctuations.

West Africa, the continent's most populous region, accounting for about 40 per cent of production, illustrates very well the variations in production consistency due to the factors that have been mentioned. Similar fluctuations are also evident in Northern, Eastern, Southern and Central Africa's production.

Figure 3.4: Regional shares (in percentages) in Africa food production value, 1991–2021

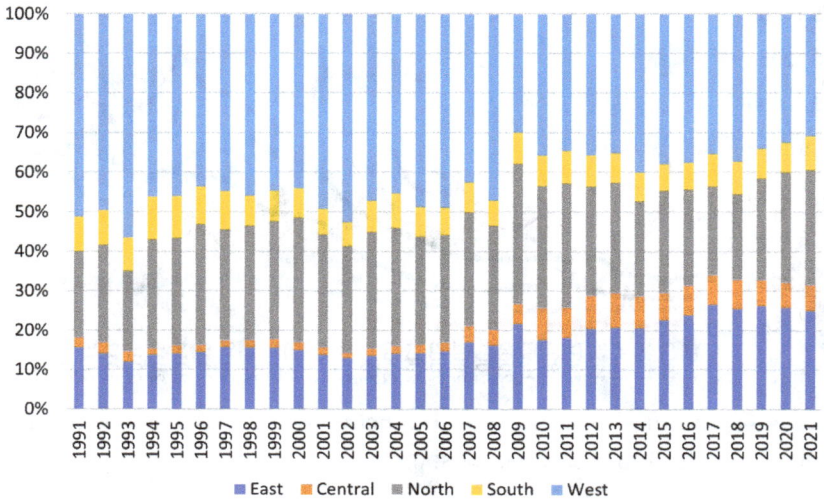

Source: Authors based on data from FAOSTAT.

3.2 The selected basic foods in relation to global production and consumption

A very revealing picture emerges from Figure 3.5. This shows Africa's global average percentage share of production and consumption of basic foods between 2010 and 2020. Keeping in mind that Africa's share of the world population was around 17 per cent during this period, this gives a broad sense of per capita distribution. Consumption outstripped production of every product except yams. This is consistent with the headline data on food security in terms of availability and stability that was noted in Chapter 1 and the food trade deficit discussed in Chapter 2.

Before looking more closely at the data on the production and consumption of these basic foods, we should note that several of these products have competing uses. An issue of interest is estimation of the food balance sheet[2] of each of the products in the basket of basic foods. Figure 3.6 shows the distribution of the main uses: (human) consumption versus feed for livestock and 'other uses'.[3] With less than half (45 per cent) of production going into consumption, yams are the produce with the lowest use for food. But this also suggests that, with yam food sufficiency attained, other uses provide an important utilisation channel.

The proportion of these products that went into livestock feed varied from 6.2 per cent of rice to about a third of maize and a fifth of cassava. In relation to 'other uses', seed accounted for 24 per cent of yam utilisation. A significant

Figure 3.5: Africa's share (in percentage) of in global food production and consumption by weight, 2010–2020

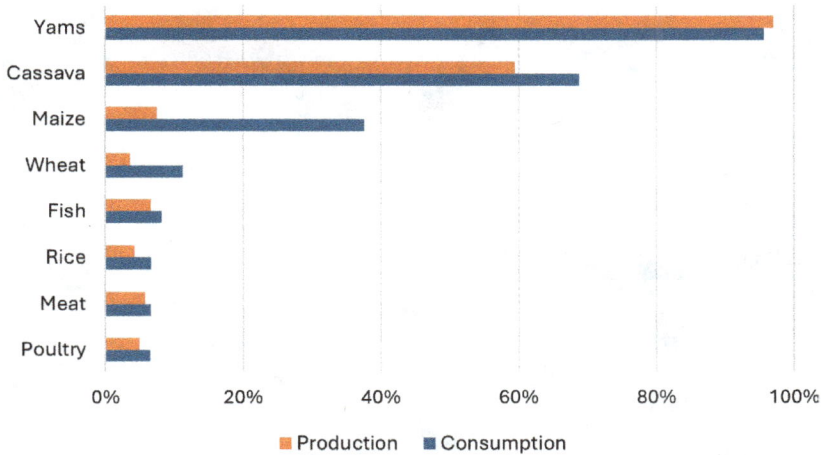

Source: Authors based on data from FAOSTAT.

Figure 3.6: How rice, wheat, cassava, maize and yams were used in Africa in 2020, percentage shares

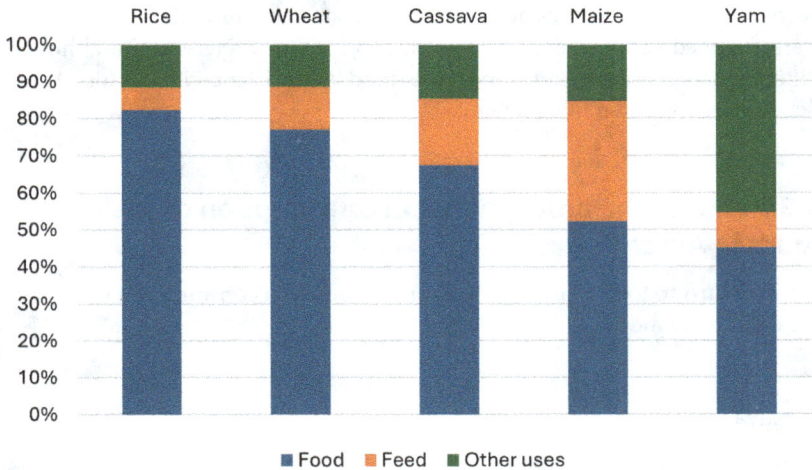

Source: Authors based on data from FAOSTAT.

proportion of cassava production, about 11 per cent, was wasted in production and processing.

Finally, Figure 3.7 presents a comparative perspective on domestic production and utilisation of the products. Apart from cassava and yams, where production and utilisation trended closely, continental utilisation of wheat, rice

Figure 3.7: Domestic production versus total utilisation, million MT, 2020

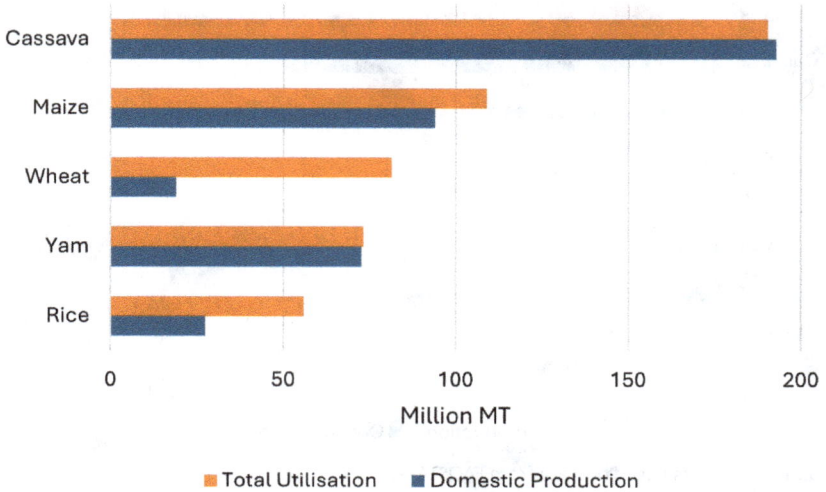

Source: Authors based on data from FAOSTAT.

and maize outstripped domestic production. Only 23 per cent, 49 per cent and 86 per cent of continental utilisation of wheat, rice and maize, respectively, were supplied through domestic production in 2020. Here again, although different data sets are being tapped, the trend is consistent with the food trade deficit that was noted in Chapter 2.

3.3 Trends in the production and consumption of basic foods in Africa

We now turn to look more closely at production and consumption of each of the eight basic foods.

Cassava

Cassava, a perennial shrub, is grown mainly by small-scale farmers for its tubers. The leaves are also edible as vegetables. Essentially a subsistence food crop, it is rich in carbohydrates, calcium, vitamins B and C and other essential minerals. It is a staple food in Central and East Africa, especially Democratic Republic of the Congo (DRC), Congo and Tanzania, and parts of West Africa.

The entire cassava plant is versatile and a boon for sustainable agriculture. Its stems, branches, leaves and tubers can be used as animal feed. This means

Figure 3.8: Cassava: Africa's production and consumption, million MT, 2010–2020

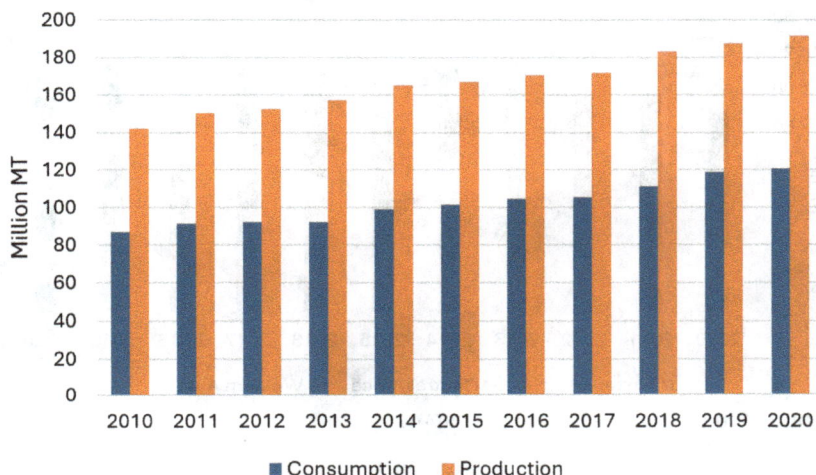

Source: Authors based on data from FAOSTAT.

that only a fraction of the land needed for cultivation of cereals and oil seed crops for animal feed is used in cassava production (Balagopalan 2001).

Industrial derivatives of cassava include starch that can be used as industrial inputs for food, medicine, cosmetics, textiles, paper, confectionery, beverages, feed, biodegradable materials, adhesives and glues, chemicals, fuel ethanol (Goodway n.d.) and so on. Cassava flour, which unlike wheat flour is gluten-free, is seen as a healthy alternative for making bread and other pastries. Cassava flour can be used as a substitute for wheat flour or mixed with wheat flour to reduce the gluten content of the final product. Nigeria's cassava flour policy (based on a 2002 presidential initiative) requires bread to contain at least 10 per cent cassava flour as a measure to reduce wheat imports and generate other economic benefits including employment along cassava value chains. A survey in Eastern Nigeria revealed that up to 97 per cent of bakeries in 2017 were using cassava flour in baking (Onyekuru et al. 2019).

Africa dominates cassava production and consumption. According to FAOSTAT, the top seven cassava growers in the world are Nigeria (60 million metric tonnes (MT)), DRC (41 million MT), Thailand (29 million MT), Ghana (22 million MT), Brazil (18.2 million MT) and Viet Nam (10.5 million MT). West, Central and Eastern Africa are the main producers, with West Africa producing about half of Africa's production. As would be expected, these regions also account for most of Africa's consumption. Figure 3.8 shows a consistent rising trend in cassava production that outstrips consumption.

As shown in Figure 3.9, all the African producing regions are net cassava producers.

Figure 3.9: Cassava production net consumption in selected subregions of Africa, million MT, 2010–2020

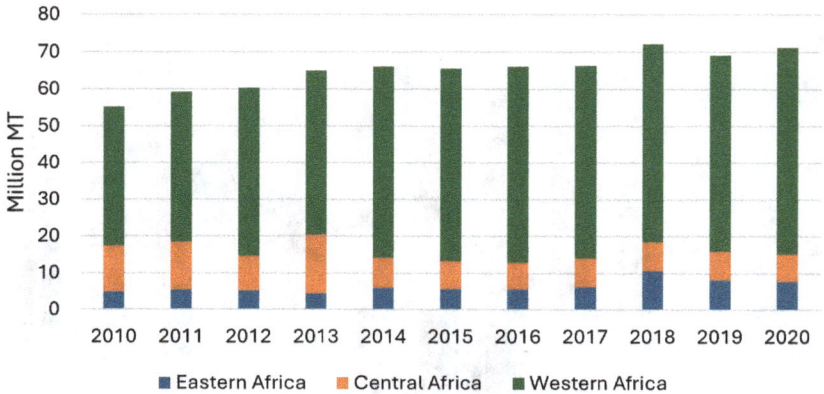

Source: Authors based on data from FAOSTAT.

Figure 3.10: Estimated cassava yields in Africa and the world, tonne/ha, 2010–2020

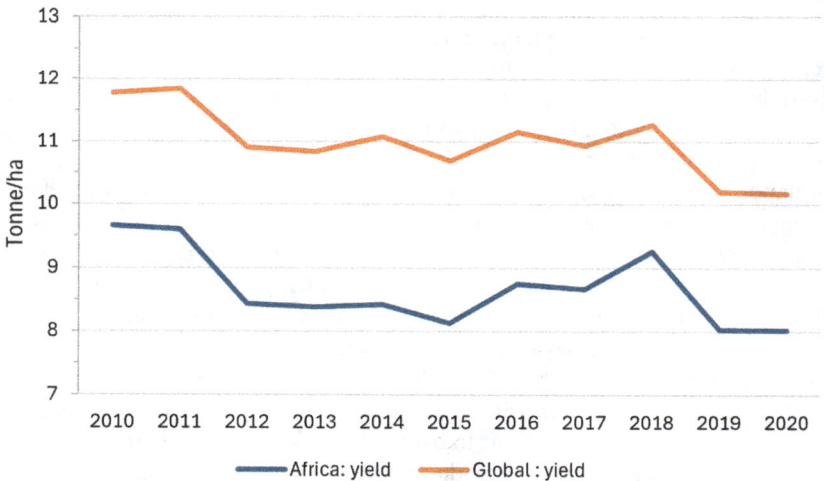

Source: Authors based on data from FAOSTAT.

Figure 3.10 provides insight into productivity challenges even for an endemic product like cassava. As can be seen, Africa's cassava yield is comparable with the global average, as the two trended together. But Africa's yield is consistently below the global average yield, suggesting room for productivity improvement. Efforts to increase cassava productivity would require inputs and practices such as improved cassava stem cuttings, adequate spacing of cassava stems during planting, improved soil preparation, application of fertilisers and irrigation. Though the crop is resilient, yields are being adversely affected by low or irregular rainfall, warmer temperatures and drier conditions that are among the features of climate change. Research suggests that cassava yield can increase sixfold when water supplied irrigation is relatively equal to that of the season's rainfall (Goodway n.d.).

With evidence to suggest that embracing modern farming methods would increase yields and adapt cassava production to climate risks, the Ibadan-based International Institute for Tropical Agriculture (IITA) is at the forefront in developing adaptation measures such as high-yielding vitamin A-fortified cassava varieties. The institute's intervention programmes include development of disease- and pest-resistant cassava varieties that are drought-resistant, early-maturing and high-yielding, with lower cyanide content.

Yams

As with cassava, the continent is important globally in yam production and consumption. And, like cassava, yams are both a subsistence crop and source of income for smallholder farmers. The tuber is rich in carbohydrates, vitamin C and some essential minerals. Nigeria and other West African countries are where this staple food is mainly grown and consumed. The crop is increasingly processed to produce yam flour, which is consumed as semolina. Unlike cassava it is not widely used as animal feed but it has industrial application as an all-purpose adhesive used in packaging and in the production of leather goods such as shoes.

The continent's yam production is strong, with a generally increasing trend to 73.2 million MT in 2020, as shown in Figure 3.11. In 2020, African countries accounted for 98 per cent of the world's 74.8 million MT yam production. Nigeria alone reported two-thirds of African production.

In terms of productivity, Africa's and global estimated yam yields are virtually the same. However, this is unsurprising given that the continent accounts for 98 per cent of global production – Africa's yam production *is* the world's yam production. Comparing African yields with those from the 2 per cent of production that occurs in other regions, estimated African yields are 82 per cent of those in Latin America and 57 per cent of those in Oceania, as of 2022.[4]

The IITA, which carries out research on selected tropical agricultural products including yams, cassava and maize, has suggested that with appropriate inputs, adaptation strategies and modern farming methods farmers should be able to achieve 20 ton/ha (Bouchene et al. 2021). This is over twice Africa's (and the world's) current average yield. Achieving higher levels of productivity is an increasingly important priority as expanding acreage under cultivation through deforestation and biodiversity loss is not sustainable.

Figure 3.11: Yam production and consumption in Africa, million MT, 2010–2020

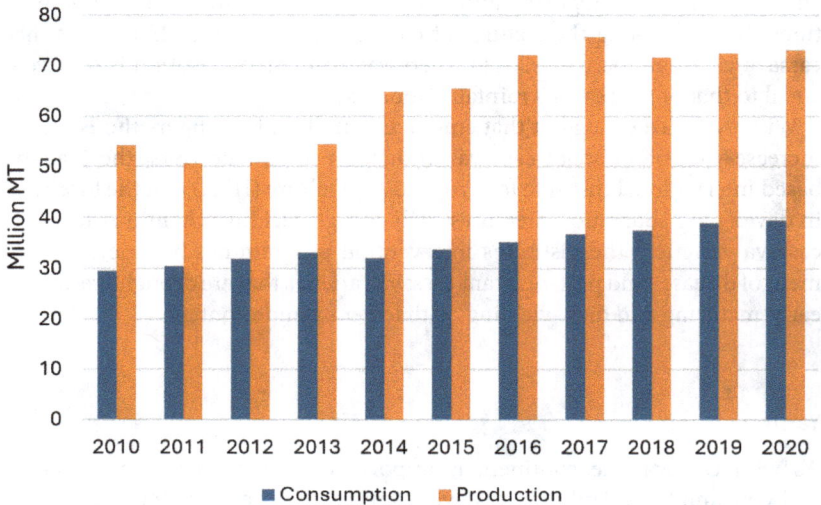

Source: Authors based on data from FAOSTAT.

Rice

Rice is an important commodity for food security and a source of dietary energy throughout the continent but especially in West Africa. It is the fourth most important source of calories in Africa, behind maize, wheat and cassava. It is a source of livelihood for more than 35 million smallholder rice farmers. A crop that can thrive in diverse ecosystems, it is cultivated in 40 out of 54 countries in Africa (Seck et al. 2012). Cultivation in Africa occurs in four ecosystems: dryland (38 per cent of the cultivated rice area), rainfed wetland (33 per cent), deep-water and mangrove swamps (9 per cent) and irrigated wetland (20 per cent) (Balasubramanian et al. 2007). However, irrigated rice production is known to be a contributor to greenhouse gas (GHG) emissions, which is a source of concern as it is estimated that land use and forestry emissions accounted for about 40 per cent of Africa's total emissions, with half of this coming specifically from agricultural activities (*AfricaRice* 2020).

In contrast to cassava and yam production, Africa's production of rice is relatively small, accounting for only 5 per cent of global production in 2020. The main African producers are in West Africa (Nigeria, Mali, Guinea, Côte d'Ivoire, Senegal and Sierra Leone), East Africa (Tanzania and Madagascar), North Africa (Egypt) and Central Africa (DRC). Each produced more than 1 million MT in 2020.

Rice consumption in Africa outstrips production, as shown in Figure 3.12. Demand for rice is estimated to be growing at more than 6 per cent a year,

Figure 3.12: Rice production and consumption in Africa, million MT, 2010–2020

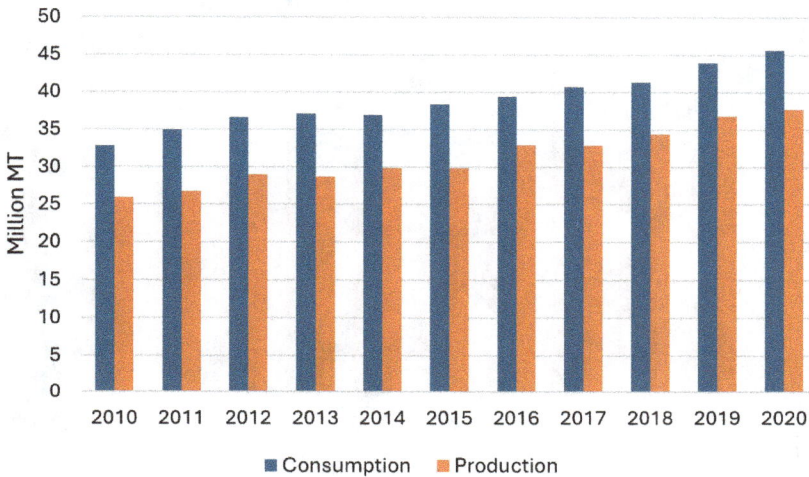

Source: Authors based on data from FAOSTAT.

making it the fastest-growing food staple in Africa. Population growth and urbanisation are the main factors driving this growth.[5] The latter has a double impact in terms of its pull factor, which drains farm communities of labour and produces a strong consumer preference for rice in urban areas. Rice is one of the products that drives Africa's status as a net food importer (see Chapter 2).

Rice yield in Africa is about half of the global average (4.69 tonnes/ha, compared to 2.40 tonnes/ha in Africa, in 2020) (Figure 3.13). But productivity is higher in Egypt, which generates a yield that is more than three times the world average, at 9.4 tonne/ha versus 3 tonne/ha. Natural endowments such as favourable climatic condition including high sunlight intensity, limited presence of pests, insects and disease, and the fertile plains of the Nile delta contribute to higher productivity in Egypt. But the success of Egyptian rice is also due to investment in well-designed irrigation systems, new short-duration, high-yielding varieties, and the use of modern production technologies in rice farming. Egypt's national Rice Campaign programme takes an adaptive approach to the monitoring of production constraints and challenges, which are promptly addressed by the various agencies that are part of the programme.

In other parts of Africa, variable rainfall in upland ecologies, drought, flood, extreme temperatures, soil quality and erosion, pests and underinvestment are among the factors accounting for low productivity. Efforts are being made to develop rice varieties such as New Rice for Africa (NERICA), with early maturity, improved yield and tolerance to major stresses, by Africa Rice Center (*AfricaRice*).

Figure 3.13: Estimated rice yields in Africa and the world, tonne/ha, 2010–2020

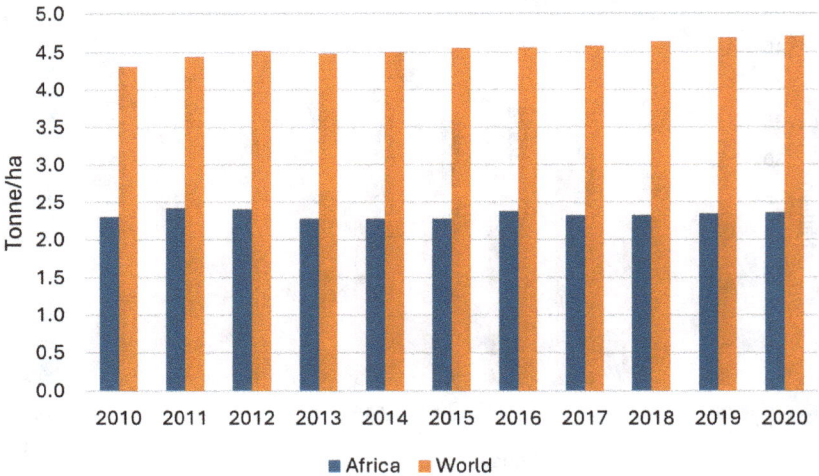

Source: Authors based on data from FAOSTAT.

Maize

Maize is an important staple food for some 300 million Africans, especially those in Central, Eastern and Southern Africa. It is a low-cost source of starch, fibre, protein, vitamins and minerals, such as magnesium, zinc, phosphorus and copper (Bathla, Jaidka and Kaur 2020). It accounts for 30–50 per cent of low-income household expenditure in Africa. Besides human consumption, maize is a major input into the manufacturing of animal feed, especially for poultry. On average, maize used in animal feeds is about a third (33.3 per cent) of the total utilisation in Africa, as depicted in Figure 3.6. Maize is also used in the manufacturing sector. For instance, about 20 per cent of maize produced in Nigeria is used in brewing beers, industrial flour, corn flakes and other confectioneries (PWC 2021, p.20).

Like rice, Africa's maize production is a relatively small share of global production. Maize is widely grown on the continent but West and Eastern Africa account for two-thirds of production. The top five African producers of maize are Nigeria, Tanzania, Ethiopia, Egypt and South Africa.[6]

Africa produces more maize than it consumes, as shown in Figure 3.14. However, the continent is a net importer of maize as the excess production over human consumption is not sufficient to meet demand for other uses of maize including feeds for livestock and industrial processing and manufacturing. For instance, in 2020, total utilisation of maize was estimated at 109.04 million MT, compared to 93.89 million MT of domestic production (Figure 3.7), leaving a balance of about 15 million MT sourced from international markets.

Figure 3.14: Maize production and consumption in Africa, million MT, 2010–2020

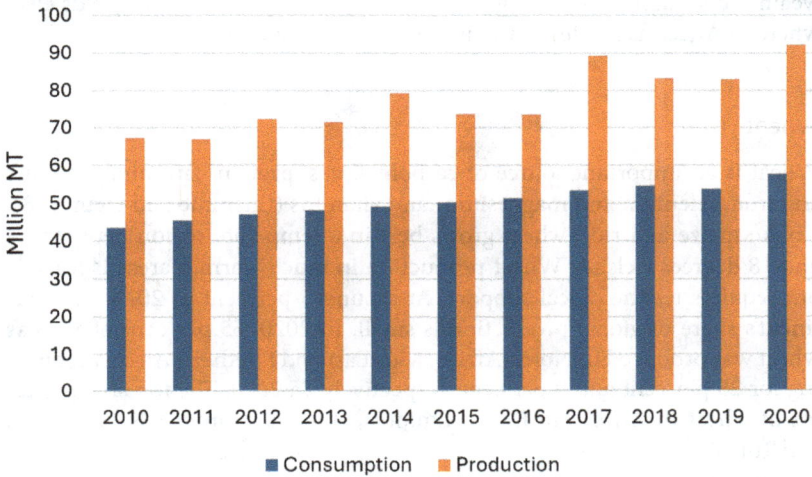

Source: Authors based on data from FAOSTAT.

Figure 3.15: Estimated maize yields in Africa and the world, tonne/ha, 2010–2020

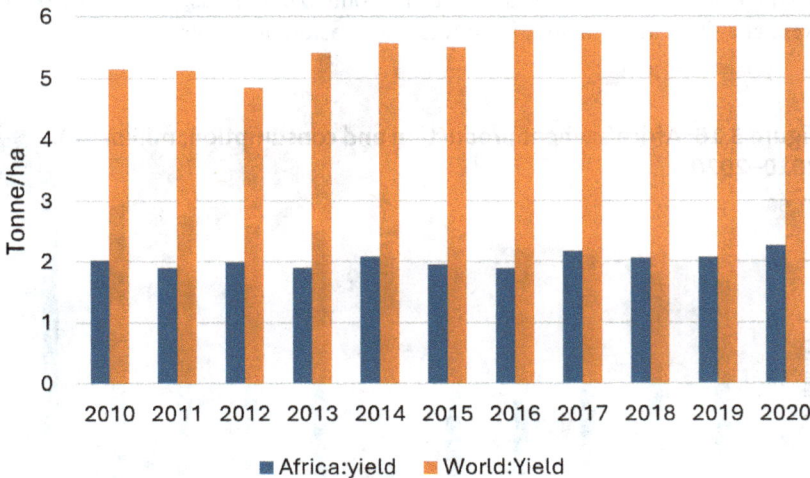

Source: Authors based on data from FAOSTAT.

However, maize yield in Africa, though increasing over the years, is less than half the world average yield (Figure 3.15). Regional differences in maize yield on the continent are wide. North Africa performed above the global average in the last four decades. In 2021, North Africa's maize productivity of 6.6 tons/ha was about 111 per cent of the global average.

There have in recent years been productivity improvements in Southern Africa, where there was an average yield of 5.0 tonne/ha in 2020. Variable weather conditions are among the factors that account for lower yields elsewhere in Africa as rainfed agricultural practices dominate.

Wheat

Wheat is an important source of carbohydrates, protein, fat, minerals (zinc and iron, selenium and magnesium) and vitamins (thiamine and vitamin B). Unlike maize and rice, wheat grows best in a temperate region between 14 and 18 degrees Celsius. Wheat production in much warmer areas is feasible but requires technological support. At around 3 per cent in 2020, the continent's share of global production is small. In 2020, 65 per cent of Africa's wheat was produced in North Africa, with East and Southern Africa accounting for 25 per cent and 9 per cent, respectively, in 2020. The top six producers of wheat in Africa are Egypt, Ethiopia, Algeria, Morocco, South Africa and Tunisia.[7]

Africa consumes more wheat than it produces, as shown in Figure 3.16. Africa's average share of global consumption was 11.2 per cent for the period between 2010 and 2020. In 2020, North Africa accounted for almost 60 per cent of Africa's wheat consumption; East and West Africa each accounted for around 15 per cent, with Southern and Central Africa making up the rest. This is why North Africa and some parts of East and West Africa were at risk from the fallout from Russia's invasion of Ukraine.

Figure 3.16: Africa's wheat production and consumption, million MT, 2010–2020

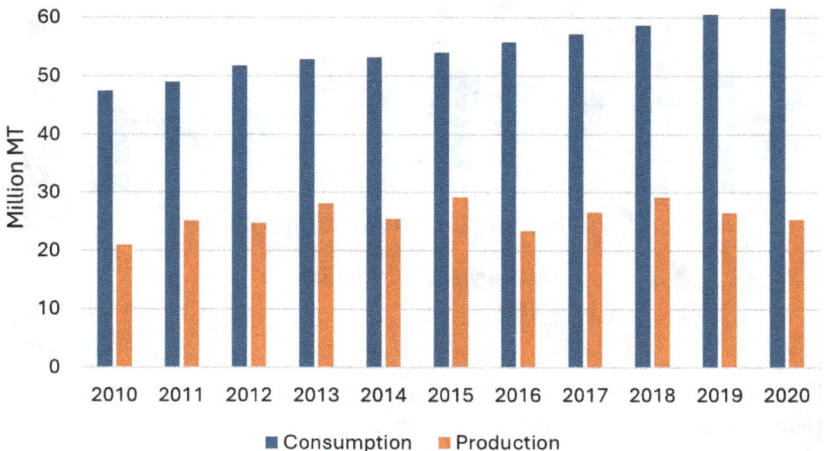

Source: Authors based on data from FAOSTAT.

Figure 3.17: Estimated wheat yields in Africa and the world, tonne/ha, 2010–2020

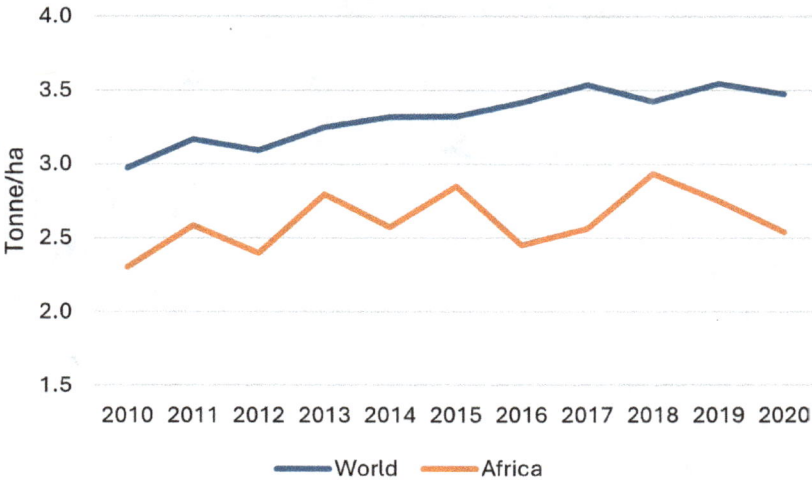

Source: Authors based on data from FAOSTAT.

Wheat productivity in Africa trends below the rest of the world, as shown in Figure 3.17. Closing the yield gap will be a challenge given that global wheat yields fell by 5.5 per cent during 1980–2010 owing to rising temperatures. Pequeno et al. (2021) simulated climate change impacts and adaptation strategies for wheat globally using new crop genetic traits. The model projected that climate change would lower global wheat production by a further 1.9 per cent by 2050. The simulation suggested that most of the negative impacts are likely to affect developing countries in the tropical regions. Africa and South Asia are expected to bear the brunt of this impact as wheat yields are projected to decline by 15 and 16 per cent, respectively, by 2050.

Meat

Meat is a source of protein, iron, vitamin B12 (as well as other B complex vitamins), zinc, selenium and phosphorus. It is obtained from different livestock types. Africa's livestock accounts for one-third of the global livestock population and about 40 per cent of agricultural GDP in Africa, ranging from 10 per cent to 80 per cent in individual countries (African Union – International Bureau for Animal Resources (AU-IBAR) 2016; Balehegn et al. 2021; Malabo Montpellier Panel 2020). However, the challenge is Africa's low output of livestock outputs as captured in the low yield. Poultry is the largest source of meat, closely followed by bovine meat. Mutton and goat meat are also important. Production of pig and other meat is smaller. (Figure 3.18). Meat of all types is the focus of this section. The next section focuses on poultry specifically.

Figure 3.18: Africa's meat by livestock, million MT, 2020

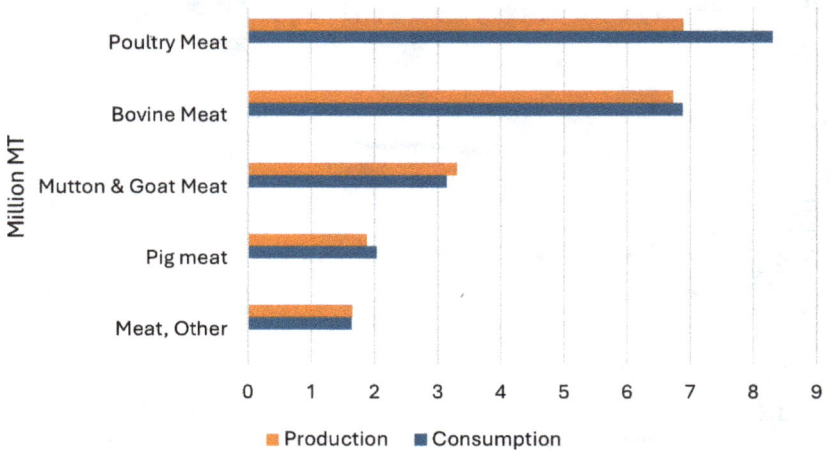

Source: Authors based on data from FAOSTAT.

Figure 3.19: Africa's production and consumption of meat, million MT, 2010–2020

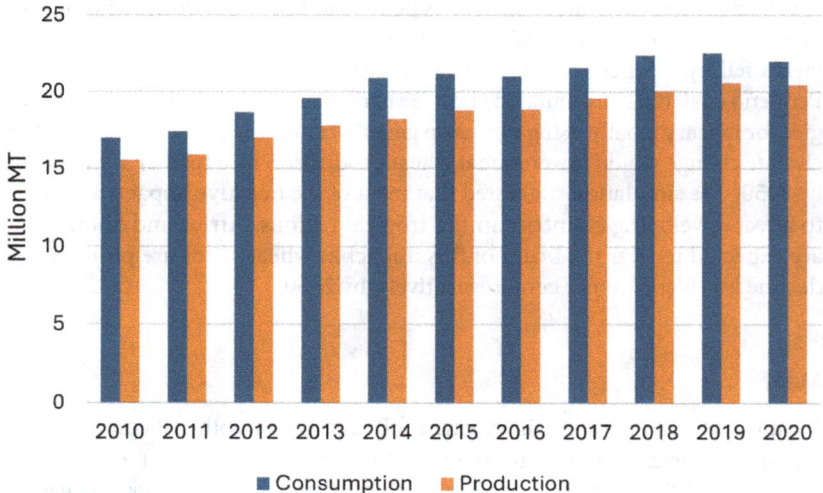

Source: Authors based on data from FAOSTAT.

Africa's total meat production was 20.5 million MT in 2020 (Figure 3.19), or about 6 per cent of the 337.8 million MT global production of meat. North and East Africa each accounted for about 27 per cent of Africa's production in 2020. West, Southern and Central Africa accounted for 20, 17 and 9 per cent, respectively.

Africa consumes more meat than it produces (Figure 3.20). This also contributes to the continent's status as a net food importer, as discussed in

Figure 3.20: Africa's net consumption of meat, thousand MT, 2010–2020

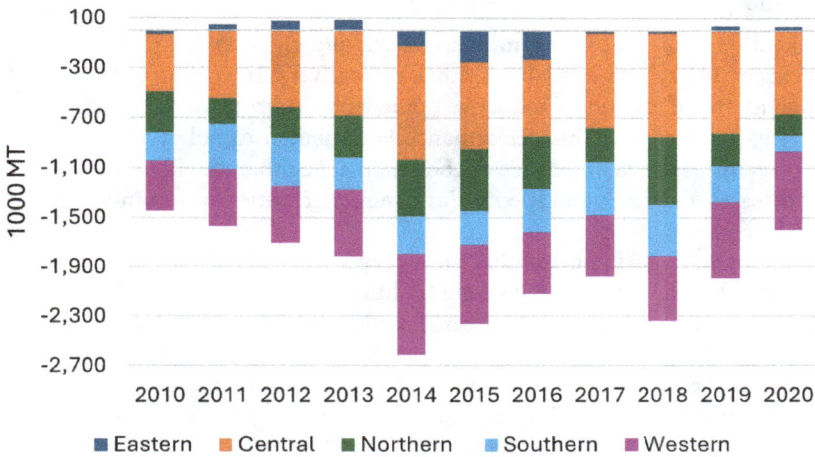

■ Eastern ■ Central ■ Northern ■ Southern ■ Western

Source: Authors based on data from FAOSTAT.

Figure 3.21: Meat food supply quantity (kg/capita/yr) by African subregion, 2010–2021

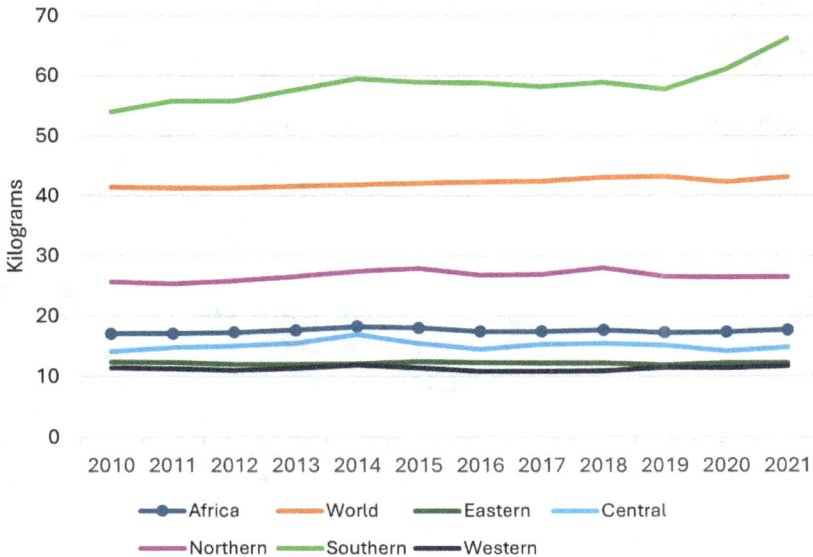

Source: Authors based on data from FAOSTAT.

Chapter 2. Africa's annual average meat consumption is consistently below 20 kg per capita and hovers around 17 to 18 kg, as shown in Figure 3.21. It is less than 50 per cent of the world average of between 41 and 43 kg. But North Africa's average per capita meat consumption of between 25 and 28 kg is above the African average, while Southern Africa's consumption of between 54 and 66 kg is not only above Africa's average but higher than the world's average.

South Africa is the most productive cattle and pig producer in Africa, with yields of 231 kg and 86 kg per animal, respectively. The country also ranked high as one of the most productive poultry producers in Africa, with a yield of 1.94 kg per bird. This performance is attributed to policy interventions especially for the maintenance of good animal health, reduction of incidence of diseases outbreaks and support for commercialisation of communal farms (Balehegn et al. 2021).

The main factors constraining meat production include (1) the quandary between husbanding lower-yielding traditional but more resilient breeds and productive exotic breeds; (2) informal markets underpinned by sociocultural systems; (3) a variety of endemic animal diseases; (4) underinvestment in facilities that support downstream and upstream production activities; (5) perennial difficulties in accessing animal feed; and (6) climate change, which is reducing the grazing land available (Rich et al. 2022).

Climate change is also driving conflicts over land between nomadic pasto-ralists and crop farmers. Rising temperatures and weather variabilities reduce the availability of feeds, forages and grazing areas, all of which contributes to lower yields (Figure 3.22). Efforts aimed at meeting projected increase in demand for meat must simultaneously address climate change risk.

Figure 3.22: Yield: beef and buffalo meat, primary, kg/animal

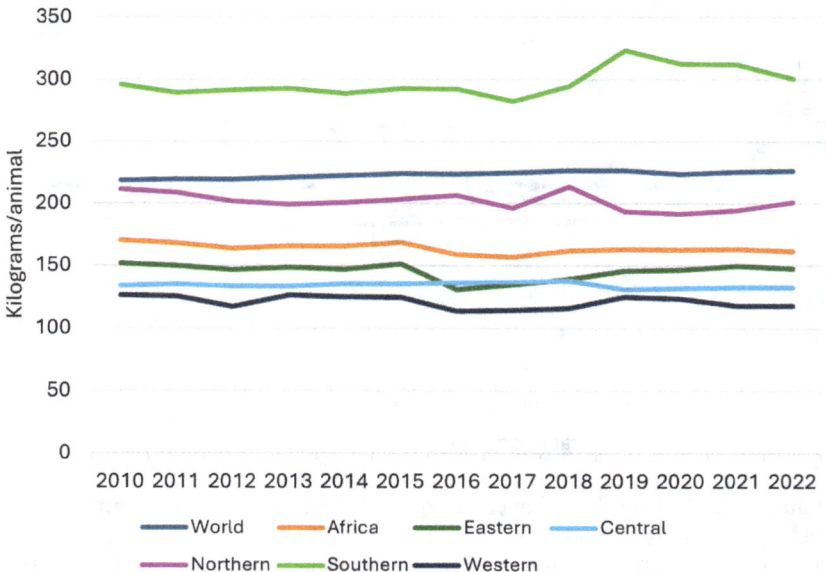

Source: Authors based on data from FAOSTAT.

African livestock has significant impacts on the environment. It is estimated that more than 70 per cent of agricultural GHG emissions in Africa come from livestock dominated by enteric methane (CH_4) emission. Methane can reduce crop yields by contributing to the formation of ground-level ozone and rising temperatures (Shindell et al. 2019). It is also estimated that the emission per unit of livestock product in Eastern Africa is four times the global average (Rich et al. 2022).

Poultry

Poultry meat, like other meat, is a good source of protein. It is rich in vitamins C and B6, iron, calcium, magnesium and cobalamin. It is also a good source of essential fatty acids. Chicken eggs have been identified to represent the lowest-cost animal source for proteins, vitamin A, iron, vitamin B12, riboflavin and choline and the second-lowest-cost source for zinc and calcium (Réhault-Godbert, Guyot and Nys 2019).

In 2020, Africa provided only 5 per cent of the world's poultry production. In terms of regional breakdown, Southern Africa accounted for 48 per cent – almost half of the continent's production – North Africa accounted for 25 per cent, West and East Africa each accounted for 12 per cent and Central Africa accounted for 2 per cent. As shown in Figures 3.23 and 3.24, Africa is a net consumer of poultry meat. But poultry meat per capita consumption, at 6 kg, is far below the global average, at

Figure 3.23: Poultry meat consumption and production in Africa, million MT, 2010–2020

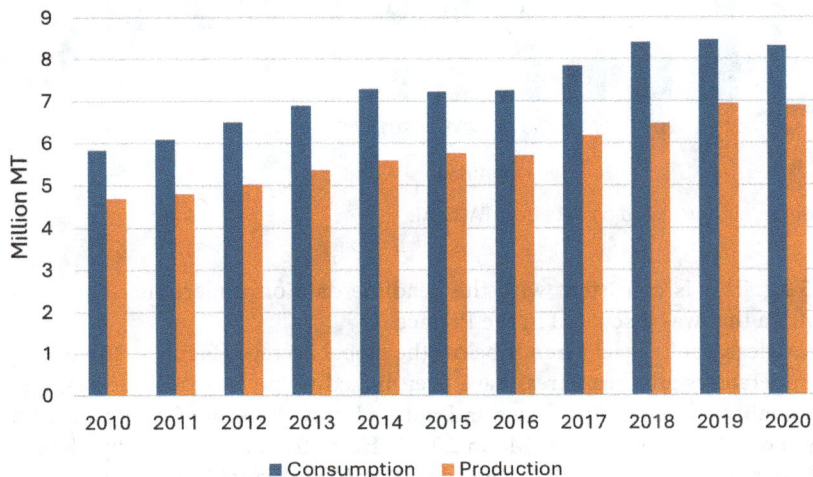

Source: Authors based on data from FAOSTAT.

Figure 3.24: Africa's poultry meat production net consumption by subregion, thousand MT, 2010–2020

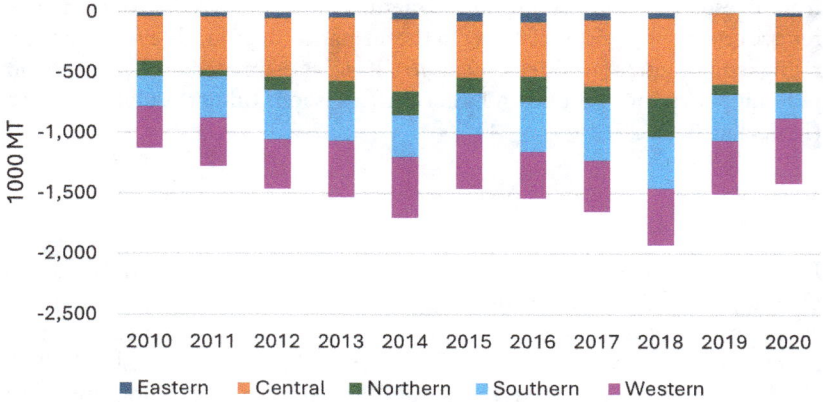

Source: Authors based on data from FAOSTAT.

Figure 3.25: Poultry yields in Africa and the world, 100g/bird, 2010–2020

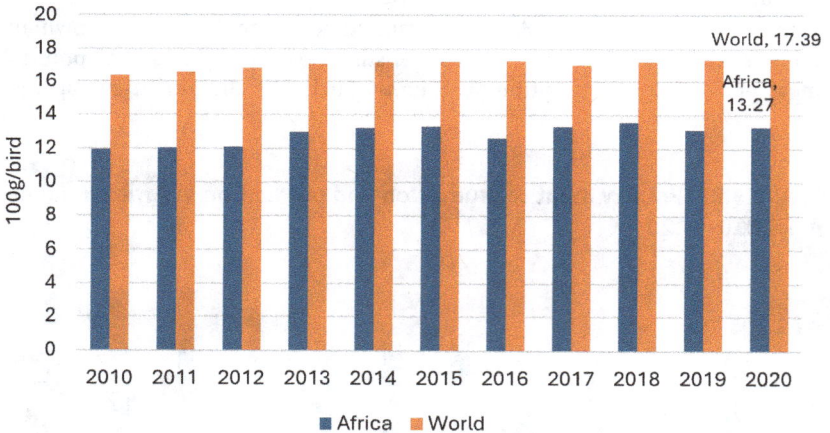

Source: Authors based on data from FAOSTAT.

15 kg. This is consistent with the headline data on undernourishment in Africa that was discussed in the Preface.

Africa's poultry yield trends below the global average (Figure 3.25). Rising temperatures and heat stress have been linked to 'poultry death losses, loss of quality and quantity of eggs and reduced growth in intensive production system' (Tabler, Wells and Moon 2021). Heat stress occasioned by climate change affects poultry production directly or indirectly in several ways. The direct effect includes a negative impact on chicken growth and productivity

(Liverpool-Tasie, Sanou and Tambo 2019), a reduction in the productive efficiency of hens and hence egg production (Mashaly et al. 2004) and a decrease
in poultry production at temperatures higher than 30°C (Ensminger, Oldfield
and Heinemann 1990). The indirect effects work through inputs to poultry
production such as quality and quantity of feed and water. It reduces feed
intake, weight gain, carcass weight and protein content (Tankson et al. 2001).

Fish

Fish is a source of high-quality and low-cost proteins. It provides essential
amino acids, omega-3 fatty acids, minerals, especially iron and zinc, and
vitamins. The three main fish production systems – marine, freshwater
and aquaculture – account for 58, 27 and 15 per cent, respectively, of Africa's production. Thirty per cent of the continent's population, approximately
200 million people, consume fish as the main animal protein source (African
Natural Resources Centre (ANRC) 2022).[8] Africa's fisheries sector employs
12.3 million people, with 6.1 million (50 per cent) employed as fishers,
5.3 million (42 per cent) as processors and 0.9 million (8 per cent) as fish
farmers. The sector is important as a source of not only nutrients but also
livelihoods (Obiero et al. 2019).

At 8 per cent in 2020, Africa's share of world fish production is relatively
small. In the regional breakdown, North Africa contributed 32 per cent of the
continent's production in 2020. This was closely followed by West Africa at
30 per cent, with East, Central and Southern Africa accounting for 19, 10 and
9 per cent, respectively. Figure 3.26 shows that Africa consumes more fish than
it produces. Consumption in West and Central Africa is mainly responsible for

Figure 3.26: Africa's fish consumption and production, million MT, 2010–2020

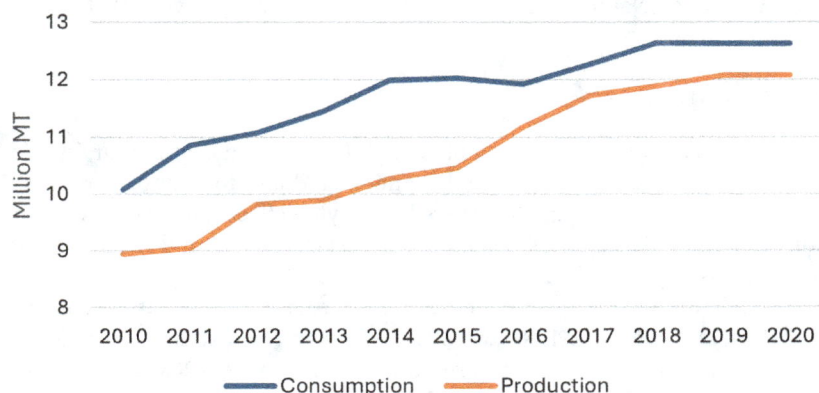

Figure 3.27: Africa's fish production net of consumption, million MT, 2010–2020

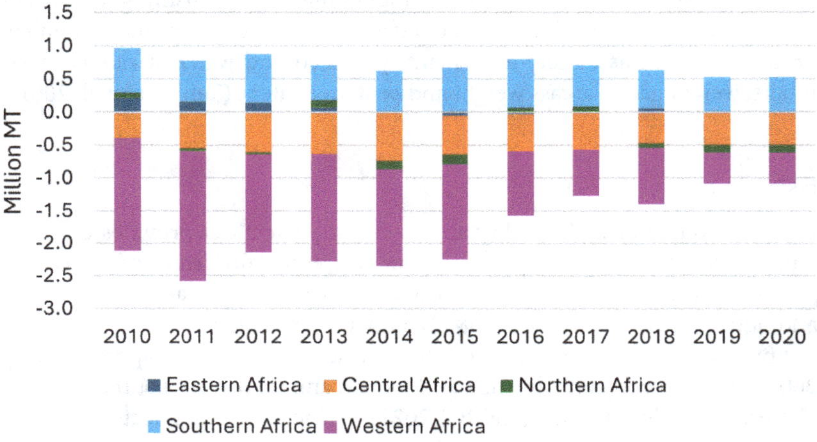

Source: Author based on data from FAOSTAT.

the deficit. Southern Africa consistently produced a significant surplus, while Eastern and North Africa produced marginal surpluses (Figure 3.27).

Challenges facing the fish sector in Africa include underinvestment in the management of fish stocks, the marine environment and freshwater habitats, illegal unregulated and unreported fishing by foreign fleets as discussed in relation to the World Trade Organization (WTO) Fisheries Agreement in Chapter 9, and rising sea temperatures. The latter has a disproportionate effect on small-scale fishing communities, which make up a large part of Africa's fisheries. As the sea temperature rises, fish stocks migrate towards colder waters. This increases pressure on small-scale fishing communities to scale up operations by investing in equipment and vessels that can go out further into the sea (African Natural Resources Centre (ANRC) 2022; Lovei 2017).

3.4 Climate risks

As noted in Chapter 2, the story of Africa's food security, agricultural trade and climate impacts is complex and cannot be reduced to a simple narrative. But it is also clear that climate change has varying effects on production and consumption of basic foods. We conclude this chapter by outlining some of these effects.

It is well known that surface temperature in Africa is rising faster than the global average. Between 1991 and 2021, Africa warmed at an average 0.3°C – that is, 0.1°C faster than in the preceding three decades (IPCC 2022; World Meteorological Organization 2023). This trend is set to continue in all IPCC scenarios, depending on the effectiveness of mitigation and adaptation measures across the world.

The consequences are clear. The sea level along African coastlines is rising more rapidly than the global mean rate, leading to coastal flooding and increased salinity of groundwater. Water stress is mounting as freshwater sources dry up. Extreme weather events have become more frequent, more severe and more diverse across the continent, with drought in East Africa, floods across much of the continent, storms in South East Africa and wildfires in North Africa (Emergency Events Database (EM-DAT), cited in World Meteorological Organization 2023).

There are several direct risks to production and consumption. These include declining crop yields under heat and water stress (FAO Regional Office for Africa 2009); shortened crop growing seasons;[9] shrinking acreage of arable land (Prowse and Braunholtz-Speight 2007; Owusu et al. 2021); higher incidence of crop pests (FAO Regional Office for Africa 2009), such as the desert locust invasion in East Africa in 2019 (Stone 2020); inundation of cropland and erosion (Müller et al. 2011); and flood-induced damage to agriculture-related infrastructure (IPCC 2014). The indirect effects are equally impactful. They include reduced labour productivity of farm workers, whether due to harsh climatic conditions[10] or to illness as vector-borne diseases proliferate (Fouque 2020) and disincentive effects leading some farmers to abandon their farms altogether (though, equally, food scarcity due to declining yields could increase food prices and result in farmers earning more).

The International Food Policy Research Institute (IFPRI) has modelled the effects of climate risks on food systems (IFPRI 2017). The study uses the International Model for Policy Analysis of Agricultural Commodities and Trade (IMPACT),[11] along with standard assumptions about changes in population, income and climate, to make a set of baseline projections on food production, consumption and trade, and the prevalence of hunger. Selected findings are summarised in Table 3.1. Overall, food production in Africa is forecast to be 8.6 per cent lower under a representative climate change scenario (relative to the counterfactual of no climate change).[12] North Africa and Central Africa

Table 3.1: Impact of climate change on food production by 2050 (as % of projected production level without climate change)

	Africa	West Africa	Central Africa	East Africa	Southern Africa	North Africa
Aggregate food	−8.6	−7.2	−11.2	−8.8	−3.2	−13.6
Cereals	−7.9	−10.0	−5.6	0.0	9.5	−19.4
Fruits and vegetables	−13.1	−10.2	−18.5	−11.6	−19.0	−15.4
Oilseeds	−6.8	−6.4	−12.5	0.0	−50.0	−14.3
Pulses	−6.7	−12.5	0.0	11.1	0.0	0.0
Roots and tubers	−7.3	−5.4	−10.0	−15.0	0.0	11.1

Source: Authors' calculations based on IFPRI (2017).

would be most severely affected, with aggregate food production declining by 13.6 per cent and 11.2 per cent, respectively. At the other end, the Southern Africa region is projected to be less affected, with agricultural output expected to decline by just 3.2 per cent by 2050.

These findings are consistent with a study by Dasgupta and Robinson (2022) of 83 countries across four regions, Africa, Americas, Asia and Europe. The study finds that climate change has been responsible for reversing some of the improvements in food security that would otherwise have been realised, with the highest impact in Africa.

Across crops, fruits and vegetables are the most vulnerable to rising temperatures. The output of fruits and vegetables is projected to fall by over 13 per cent. Cereals are estimated to decline by 7.9 per cent.

Wheat, as discussed in this chapter, is especially sensitive to warming. As noted above, at current warming trends, by 2050, wheat yields in Africa are expected to fall by 15 per cent from levels over the period 1980 to 2010 (Pequeno et al. 2021).

Country-level studies show similar dramatic effects on crop yield or output. Empirical studies for Angola, Lesotho and Mozambique (Hunter et al. 2020), Cameroon (Molua 2008) and South Africa (Calzadilla et al. 2014) illustrate the adverse effects of climate change on agricultural yield and food security, although the outcomes vary across climate change scenarios, sectors and countries.

As noted above, meat and poultry are also at risk. In a global study, Thornton et al. (2021) project that livestock species in many parts of the tropics and some temperate zones would come under extreme heat stress by 2100, challenging the viability of outdoor livestock production systems. Anecdotal evidence of farmers switching to more drought-tolerant livestock species or breeds in Southern Africa confirms that climate change is already impacting livestock in the region (Dzama 2016).

Climate change is also connected to increasing pests. The 2019–2020 locust infestations in Ethiopia, Kenya and Somalia, which affected 1.25 million hectares of land, could be a warning of what is to come. It was also noted in this chapter that rising sea levels, temperatures and acidity are altering ocean, coastal and inland waterbodies ecosystems and displacing fish stocks. A World Bank study estimates that West and Central African countries such as Côte d'Ivoire, Liberia, DRC, Gabon and Sao Tome and Principe could see their maximum catch potential decrease by 30 per cent or more by 2050 (World Bank 2019). In East Africa, ocean warming has already destroyed parts of the coral reef (Lovei 2017).

Adaptation and mitigation

Emerging adaptation and mitigation efforts were illustrated in our discussion of the basic foods. Promising innovations are being developed to support adaptation and resilience-building in food systems while increasing agricultural productivity. These strategies could be classified into three groups:

autonomous adaptation, adoption of climate-smart technologies and trans-formational changes, and behavioural changes, including through trade poli-cies and shifts in diet (Nhamo et al. 2019).

Autonomous adaptation involves incremental approaches based on learning-by-doing at the level of individual farmers or agricultural traders (Vermeulen et al. 2018), such as altering inputs (seed varieties, fertilisers) to improve resilience to heat or drought; changing the timing or location of cropping activities; using water more effectively and managing soil mois-ture; and adopting better pest, disease or weed management practices (IPCC 2007). However, incremental adaptation may be insufficient to address rapid or unexpected climate-induced shifts in agricultural production. These changes call for climate-smart technologies like altering the resource (land, labour, capital, technology) mix or the outputs and outcomes (the types and amounts of agricultural production) (Vermeulen et al. 2018). Trade can also help countries adapt to extreme weather events that destroy crops and reduce food supply. As a short-term palliative, food imports can make up for the shortfall of production or any nutrient gap (FAO 2018).

A review of nationally determined contributions (NDCs) submitted by Afri-can countries to the UN Framework Convention on Climate Change reveals a spectrum of adaptation and mitigation measures that are related to agricul-ture.[13] Most of the adaptation measures are in the domain of autonomous adap-tation and relate to four common themes: the implementation of early-warning systems (e.g. Angola, Namibia, Zambia, Zimbabwe); water management (e.g. Botswana, Eswatini, Madagascar); crop management including diversification and adoption of drought-resistant varieties (e.g. Angola, Burkina Faso, Mada-gascar, Sao Tome and Principe, Zambia, Zimbabwe); and infrastructure devel-opment, which is included in almost every NDC. In relation to mitigation, the renewable energy sector is strongly prioritised in the NDCs.

However, these transitions are expensive, which is why finance has been such an important part of global climate discussions. It is estimated that US$2.8 trillion will be needed from 2020 to 2030 to implement NDCs in Africa, as dis-cussed in Chapter 4 in the context of resources required to support agricultural development. Mitigation accounts for two-thirds of reported climate finance needs for the period 2020–2030, distributed across the following four sectors: transport (58 per cent), energy (24 per cent), industry (7 per cent) and agricul-ture, forestry, and other land use (AFOLU, 9 per cent). Adaptation represents 24 per cent of total climate finance, even though, for Africa, adaptation, rather than mitigation, remains the dominant priority. This is why the African climate negotiators placed strong emphasis on adaptation finance at the 2023 Confer-ence of the Parties (COP 28). African governments have committed to con-tributing 10 per cent of the total cost of climate action. This means that US$2.5 trillion (or an average of US$250 billion annually) needs to be mobilised exter-nally. In 2020, Africa's climate finance flows, both domestic and international, came to only US$30 billion, or about 12 per cent of estimated requirements. These issues are discussed in greater detail in Chapter 4.

Summary

This chapter has reviewed the main trends in the production and consumption of eight basic foods, detailing how production of these foods in Africa have generally underperformed in relation to global output despite nominal growth in production. Food production growth rates have trended below population growth. Consumption has outstripped production of every product except yams. These insights brought into view the underlying dynamics not only on Africa's status as a net food importer but also on the headline data on severe food insecurity and undernourishment. This analysis also highlights the regional variations in production and consumption. Similarly, distinctions were made between different uses to which five of the basic foods – rice, wheat, cassava, maize and yams – are put.

Cassava and yams are tropical crops that are competitively produced in the African regions concerned. Unlike rice, wheat and maize or beef and poultry, which benefit from significant subsidies in richer countries with trade-distorting effect, the comparative advantage of the African cassava- and yam-producing countries remains dominant. The paradoxical effect of subsidies that disincentivise production elsewhere while also making food more widely available are further discussed in Chapter 9 on the WTO legal framework.

This analysis has further highlighted the varying effects of climate change on production of basic foods and in particular rising temperatures and extreme weather variations. African countries have prioritised a variety of adaptation and mitigation measures in their NDCs, but financing continues to be contentious. At the same time, agricultural activities contribute to Africa's total emissions. For example, enteric fermentation of ruminant livestock and irrigated rice farming practices are significant contributors to methane emissions and other GHGs. More broadly, land as a central and important input in the agricultural value chain is simultaneously a source and a sink of carbon emissions. For cassava, yams, rice, maize and wheat, the result of research into adaptation strategies is being applied, although it is also clear that underinvestment in these food systems is a constraint. In the case of livestock, adaptation measures include better grazing land management, improved manure management, higher-quality feed, use of breeds and genetic improvement. Technology and infrastructure feature prominently among mitigation measures.

Notes

[1] Nevertheless, movement away from agriculture could explain why food production is causing the continent's population (and economic) growth to lag. The share of the African labour force employed in agriculture shrank from 2011 to 2019, but food prices and yields have been increasing recently (Roser 2023; Okou, Spray and Unsal 2022). This could suggest that food production is lagging economic growth because fewer

Africans are engaged in it than before, but it is still maintaining its share in GDP because of price rises.

[2] We can think of the food balance sheet in terms of the total quantity of foodstuffs produced in a country added to the total quantity imported and adjusted to any change in stocks that may have occurred since the beginning of the reference period. This gives the supply available during that period. On the utilisation side, a distinction could be made between the quantities exported, fed to livestock, used for seed, put to manufacture for food use and non-food uses, losses during storage and transportation, and food supplies available for human consumption.

[3] Other uses include seed for the next cycle of production, non-food (industrial) uses, processed food, and losses during storage, transportation and processing.

[4] In Asia, the picture is mixed. Africa has higher yields than Bhutan and the Philippines, slightly lower yields than Malaysia (Africa's yield is 95 per cent of Malaysia's) and much lower yields than Japan (which has almost three times Africa's yield). However, Japan's yam production is dominated by the Chinese yam, whereas the rest of the world mainly grows the White Guinea yam. This difference in which yams are grown may, at least in part, explain the difference in yields (Hamaoka et al. 2022; IITA n.d.).

[5] Urbanisation apparently contributes to rice demand by providing better-paid income-earning opportunities, increasing the opportunity cost of time spent on food production. Rice is quick and easy to prepare relative to staple foods, which is why urbanisation is contributing to increased demand for rice (Rutsaert, Demont and Verbeke 2013). For more perspective on rice research and development see: https://www.africarice.org.

[6] According to FAOSTAT they produced 15.3, 12, 10, 7.6 and 6.7 million MT of maize, respectively, in 2020.

[7] Other producers include Sudan, Kenya, Zambia, Zimbabwe and Libya.

[8] The NEPAD agency Fisheries and Aquaculture Programme presents a comprehensive analysis of issues in the sector: (*The NEPAD Agency Fisheries and Aquaculture Programme* 2022).

[9] Ofori, Cobbina and Obiri (2021) estimate that the growing period may be reduced by an average of 20 per cent, resulting in a 40 per cent drop in cereal yields.

[10] Rohat et al. (2019) project that the number of people exposed to 'dangerous heat' in Africa could increase by a multiple of 20–52, reaching 86–217 billion person-days per year by the 2090s.

[11] The IMPACT model is an interlinked system of climate, water, crop and economic models designed to explore the effects of changes in climate and other factors on agricultural production, trade and food security (International Food Policy Research Institute 2017).

[12] The climate change scenario assumes the IPCC's representative concentration pathway (RCP) 8.5 and the Hadley Centre Global Environment Model version 2-Earth System general circulation model.

[13] The NDCs are available from the NDC registry: https://unfccc.int/NDCREG

References

African Natural Resources Centre (2022) *The Future of Marine Fisheries in the African Blue Economy*, Abidjan: African Development Bank Group.

African Union – Inter-African Bureau for Animal Resources (AU-IBAR) (2016) 'Livestock Policy Landscape in Africa: A Review'. http://www.au-ibar.org/component/jdownloads/finish/36-vet-gov/2712-livestock-policy-landscape-in-africa-a-review

AfricaRice (2020) 'Why Rice Matters for Africa' *AfricaRice*. https://perma.cc/JZF6-KBA4

Balagopalan, C. (2001) 'Cassava Utilization in Food, Feed and Industry', in *Cassava: Biology, production and utilization*. CABI Wallingford UK, pp.301–318. https://perma.cc/JX4T-TTGY

Balasubramanian, V., M. Sie, R. J. Hijmans and K. Otsuka (2007) 'Increasing Rice Production in Sub-Saharan Africa: Challenges and Opportunities', in D. L. Sparks (ed.) *Advances in Agronomy*. Academic Press, pp.55–133. https://doi.org/10.1016/S0065-2113(06)94002-4

Balehegn, Mulubrhan; Kebreab, Ermias; Tolera, Adugna; Hunt, Sarah; Erickson, Polly, Crane, Todd A.; and Adesogan, Adegbola T (2021) 'Livestock Sustainability Research in Africa with a Focus on the Environment', *Animal Frontiers*, vol. 11, no. 4, pp.47–56. https://doi.org/10.1093/af/vfab034

Bathla, Shikha; Jaidka, Manpreet; and Kaur, Ramanjit (2020) 'Nutritive Value', in *Maize – Production and Use*, IntechOpen. https://doi.org/10.5772/intechopen.88963

Bouchene, Lyes; Cassim, Ziyad; Engel, Hauke; Jayaram, Kartik; and Kendall, Adam (2021) *Green Africa: A Growth and Resilience Agenda for the Continent How the Global Climate Agenda Creates Opportunities for Africa to Build Resilience, Catalyze Sustainable Growth, and Contribute to the Net-Zero Transition*. McKinsey. https://perma.cc/79YQ-46TA

Calzadilla, Alvaro; Zhu, Tingju; Rehdanz, Katrin; Tol, Richard S. J.; and Ringler, Claudia (2014) 'Climate Change and Agriculture: Impacts and Adaptation Options in South Africa', *Water Resources and Economics,* vol. 5, pp.24–48. https://doi.org/10.1016/j.wre.2014.03.001

Dasgupta, Surajeet; and Robinson, Elizabeth J. Z. (2022) 'Attributing Changes in Food Insecurity to a Changing Climate', *Scientific Reports,* vol. 12, 4709. https://doi.org/10.1038/s41598-022-08696-x

Dzama, Kennedy (2016) 'Is the Livestock Sector in Southern Africa Prepared for Climate Change?', Policy Briefing 153. Johannesburg: South African Institute of International Affairs (SAIIA). https://perma.cc/5SXV-6KXZ

Ensminger, M. E.; Oldfield, J. E.; and Heinemann, W. W. (1990) *Feeds and Nutrition: Formerly, Feeds and Nutrition, Complete.* 2nd edition, Clovis, CA: Ensminger.

FAO Regional Office for Africa (2009) 'Climate Change in Africa: The Threat to Agriculture', FAO. https://perma.cc/7J5L-JKNA

FAO (2018) *The State of Agricultural Commodity Markets 2018. Agricultural Trade, Climate Change and Food Security,* Rome: FAO.

Food and Agriculture Organization of the United Nations (n.d.) 'FAOSTAT'. https://perma.cc/5Y7P-WM75

Fouque, Florence (2020) 'Climate Change Impacts the Transmission of Vector-Borne Diseases', *Research Outreach.* https://perma.cc/82W6-62Z9

Goodway (n.d.) 'Nine Uses and Industry Applications of Cassava Starch', Goodway. https://perma.cc/MM5M-GUAJ

Hamaoka, Norimitsu; Moriyama, Takahito; Taniguchi, Takatoshi; Suriyasak, Chetphilin; and Ishibashi, Yushi (2022) 'Identification of Phenotypic Traits Associated with Tuber Yield Performance in Non-Staking Cultivation of Water Yam (Dioscorea alata L.)', *Agronomy,* vol. 12, no. 10. https://doi.org/10.3390/agronomy12102323

Hunter, R.; Crespo, O.; Coldrey, K; Cronin, K; and New, M. (2020) 'Climate Change and Future Crop Suitability in Lesotho', Research Highlights, University of Cape Town and International Fund for Agricultural Development (IFAD). https://perma.cc/32RV-FSGE

IITA (n.d.) *Yam (Dioscorea species),* IITA. https://perma.cc/HMA5-UXTA

International Food Policy Research Institute [IFPRI] (2017) '2017 Global Food Policy Report', International Food Policy Research Institute. https://hdl.handle.net/10568/141778

IPCC (2007) *Climate Change 2007: Synthesis Report. Contribution of Working Groups I, II and III to the Fourth Assessment Report of the*

Intergovernmental Panel on Climate Change [Core Writing Team, Pachauri, R.K and Reisinger, A. (eds.)], Geneva: IPCC. https://perma.cc/D7E5-4J3B

IPCC (2014) *Climate Change 2014: Synthesis Report. Contribution of Working Groups I, II and III to the Fifth Assessment Report of the Intergovernmental Panel on Climate Change [Core Writing Team, R.K. Pachauri and L.A. Meyer (eds.)]*, Geneva: IPCC. https://perma.cc/499Z-CSGS

IPCC (2023) *Climate Change 2022 – Impacts, Adaptation and Vulnerability: Working Group II Contribution to the Sixth Assessment Report of the Intergovernmental Panel on Climate Change*, 1st edition, Cambridge and New York: Cambridge University Press. https://doi.org/10.1017/9781009325844

Liverpool-Tasie, Lenis S. O.; Sanou, Awa; and Tambo, Justice A. (2019) 'Climate Change Adaptation among Poultry Farmers: Evidence from Nigeria', *Climate Change*, vol. 157, pp.527–44.

Lovei, Magda (2017) 'Climate Impacts on African Fisheries: The Imperative to Understand and Act', *World Bank Blogs*, 11 November. https://blogs.worldbank.org/en/nasikiliza/climate-impacts-on-african -fisheries-the-imperative-to-understand-and-act

Malabo Montpellier Panel (2020) 'Meat, Milk and More: Policy Innovations to Shepherd Inclusive and Sustainable Livestock Systems in Africa', International Food Policy Research Institute. https://doi.org/10.2499/9780896293861

Mashaly, M. M.; Hendricks, G. L. 3rd; Kalama, M. A.; Gehad, A. E.; Abbas, A. O.; and Patterson, P. H. (2004) 'Effect of Heat Stress on Production Parameters and Immune Responses of Commercial Laying Hens', *Poultry Science*, vol. 83, no. 6, pp.889–94. https://www.sciencedirect.com/science/article/pii/S0032579119426278 ?via%3Dihub

Molua, Ernest L. (2008) 'Turning Up the Heat on African Agriculture: The Impact of Climate Change on Cameroon's Agriculture', *African Journal of Agricultural and Resource Economics*, vol. 2, no. 1, pp.45–64.

Müller, Christoph; Cramer, Wolfgang; Hare, William L.; and Lotze-Campen, Hermann (2011) 'Climate Change Risks for African Agriculture', *Proceedings of the National Academy of Sciences*, vol. 108, no. 11, pp.4313–15.

The NEPAD Agency Fisheries and Aquaculture Programme (2022) 'AUDA-NEPAD African Union Development Agency'. https://perma.cc/UVZ2-XTHX

Nhamo, Luxon; Matchaya, Greenwell; Mabhaudhi, Tafadzwanashe; Nhlengethwa, Sibusiso; Nhemachena, Charles and Mpandeli, Sylvester (2019) 'Cereal Production Trends under Climate Change: Impacts and

Adaptation Strategies in Southern Africa', *Agriculture*, vol. 9, no. 2, p.30. https://doi.org/10.3390/agriculture9020030

Obiero, Kevin; Meulenbroek, Paul; Drexler, Silke; Dagne, Adamneh; Akoll, Peter; Odong, Robinson; Kaunda-Arara, Boaz; and Waidbacher, Herwig (2019) 'The Contribution of Fish to Food and Nutrition Security in Eastern Africa: Emerging Trends and Future Outlooks', *Sustainability*, vol. 11, no. 6, p.1636. https://doi.org/10.3390/su11061636

Ofori, Samuel A.; Cobbina, Samuel J.; and Obiri, Samuel (2021) 'Climate Change, Land, Water, and Food Security: Perspectives from Sub-Saharan Africa', *Frontiers in Sustainable Food Systems*, 5. https://doi.org/10.3389/fsufs.2021.680924

Okou, Cedric; Spray, John; and Unsal, D. Filiz (2022) 'Africa Food Prices Are Soaring Amid High Import Reliance', *IMF blog*, 27 September. https://perma.cc/N45Q-WHP3

Onyekuru, N. A. et al. (2019) 'Effectiveness of Nigeria Policy on Substitution of Wheat for Cassava Flour in Bakery Products', *Nigerian Agricultural Policy Research Journal*, vol. 6, no. 1, pp.62–72. https://perma.cc/WNZ2-PHLF

Owusu, Victor; Ma, Wanglin; Emuah, Dorcas; and Alan Renwick[b] (2021) 'Perceptions and Vulnerability of Farming Households to Climate Change in Three Agro-ecological Zones of Ghana', *Journal of Cleaner Production*, vol. 293, p.126134. https://doi.org/10.1016/j.jclepro.2021.126154

Pequeno, Diego N. L.; Hernández-Ochoa, Ixchel M; Reynolds, Matthew; Sonder, Kai; MoleroMilan, Anabel; Robertson, Richard D; Lopes, Marta S; and Xiong, Wei, et al. (2021) 'Climate Impact and Adaptation to Heat and Drought Stress of Regional and Global Wheat Production', *Environmental Research Letters*, vol. 16, no. 5, p.054070. https://doi.org/10.1088/1748-9326/abd970

Prowse, Martin; and Braunholtz-Speight, Tim (2007) 'The First Millennium Development Goal, Agriculture and Climate Change', *ODI Opinion*, 85. https://perma.cc/69Y9-REX5

PWC (2021) *Positioning Nigeria as Africa's leader in maize production for AfCFTA*. PricewaterhouseCoopers Nigeria. https://perma.cc/KBA7-CXQ6

Réhault-Godbert, Sophie; Guyot, Nicolas; and Nys, Yves (2019) 'The Golden Egg: Nutritional Value, Bioactivities, and Emerging Benefits for Human Health', *Nutrients*, vol. 11, no. 3, p.684. https://doi.org/10.3390/nu11030684

Rich, Karl M.; Schaefer, K. Aleks; Thapa, Bhawna; Hagerman, Amy D.; and Shear, Hannah E. (2022) 'An Overview of Meat Processing in Africa', in Jenane, Chakib, Ulimwengu, John M. and Tadesse, Getaw (eds) *2022*

Annual Trends and Outlook Report: Agrifood Processing Strategies for Successful Food Systems Transformation in Africa. Washington, DC: International Food Policy Research Institute, pp.33–43. https://perma.cc/Z5KA-HUSE

Rohat, Guillaume; Flacke, Johannes; Dosio, Alessandro; Dao, Hy; and van Maarseveen, Martin (2019) 'Projections of Human Exposure to Dangerous Heat in African Cities Under Multiple Socioeconomic and Climate Scenarios', *Earth's Future*, vol. 7, no. 5, pp.528–46. https://doi.org/10.1029/2018EF001020

Roser, Max (2023) 'Employment in Agriculture', *Our World in Data* [Preprint]. https://perma.cc/BJ2T-AUWT

Rutsaert, P.; Demont, M.; and Verbeke, W. (2013) 'Consumer Preferences for Rice in Africa', in M. C. S. Wopereis et al. (eds) *Realizing Africa's Rice Promise*, CABI Publishing, pp.293–301. https://doi.org/10.1079/9781845938123.0294

Seck, Papa Abdoulaye, Aliou Diagne, Samarendu Mohanty and Marco C.S. Worpereis (2012) 'Crops that feed the world 7: Rice', *Food Security*, 4(1), pp.7–24. https://doi.org/10.1007/s12571-012-0168-1

Secretariat of the Paris Agreement (n.d.) *NDC Registry, United Nations Climate Change | Nationally Determined Contributions Registry.* https://unfccc.int/NDCREG

Shindell, Drew et al. (2019) 'Spatial Patterns of Crop Yield Change by Emitted Pollutant', *Earth's Future*, vol. 7, no. 2, pp.101–12. https://doi.org/10.1029/2018EF001030

Stone, Madeleine (2020) 'A Plague of Locusts Has Descended on East Africa. Climate Change May Be to Blame', *National Geographic*, 14 February. https://perma.cc/96RB-X25L

Tabler, Tom; Wells, Jessica; and Moon, Jonathan (2021) 'Poultry Industry's Contribution to Greenhouse Gas Emissions', *Mississippi State University Extension Service Publication*, vol. 3634. https://perma.cc/3FHU-NJKB

Tankson, J. D.; Vizzier-Thaxton, Y.; Thaxton, J. P.; May, J. D.; and Cameron, J. A. (2001) 'Stress and Nutritional Quality of Broilers', *Poultry Science*, vol. 80, no. 9, pp.1384–89. https://doi.org/10.1093/ps/80.9.1384

Thornton, Philip; Nelson, Gerald; *Mayberry*, Dianne; and Herrero, Mario (2021). 'Increases in Extreme Heat Stress in Domesticated Livestock Species during the Twenty-First Century', *Global Change Biology*, vol. 27, pp.5762–72. https://doi.org/10.1111/gcb.15825

Vermeulen, S. J.; Dinesh, D.; Howden, S. M.; Cramer, L.; and Thornton, P. K. (2018) 'Transformation in Practice: A Review of Empirical Cases of

Transformational Adaptation in Agriculture under Climate Change', *Frontiers in Sustainable Food Systems*, vol. 2, no. 65, pp.1–17. https://doi.org/10.3389/fsufs.2018.00065

World Bank (2019) *Climate Change and Marine Fisheries in Africa: Assessing Vulnerability and Strengthening Adaptation Capacity.* Washington, DC: The World Bank. https://hdl.handle.net/10986/33315

World Bank (n.d.) 'World Bank Open Data'. https://data.worldbank.org

World Meteorological Organization (2023) *State of the Climate in Africa 2022.* 1330. World Meteorological Organization. https://perma.cc/2Z97-AYHM

4. Policy, resources, actors and capacities

Vinaye Dey Ancharaz

Agricultural policies are important determinants of food security outcomes. Finance, investment, institutions, actors and capacities interact with policies in playing a key role in resource allocation along the food value chain, from production to consumption, from supply to demand. Such policies are most effective when they are evidence-based and adapt to changing realities.

To coordinate agricultural policies across the continent, a common framework for such initiatives has long been an objective of the member states of the African Union and its predecessor, the Organisation of African Unity. The 2003 African Union (AU) Comprehensive Africa Agriculture Development Programme (CAADP) compact responded to this collective aspiration. This chapter discusses how effective Africa's agricultural policies are, as well as how far countries have implemented the framework set out in CAADP.

4.1 Agricultural policy and implementation

The effort to provide a continental policy framework on agriculture can be traced back to the 1980 Lagos Plan of Action, which recognised the need for the sector to be prioritised for economic development and poverty reduction. However, this effort fell short of proposing a continental strategy for the agricultural sector and was overshadowed by the structural adjustment programmes of the era (Badiane, Collins and Ulimwengu 2020). With the establishment of the African Union in 2001 and the reorientation of its development priorities through the New Partnership for Africa's Development (NEPAD), agriculture came back into focus. In 2003, the CAADP compact was agreed. The AU's Agenda 2063 revalidated CAADP in 2013 as the continent's strategy for achieving agricultural development and food security.

How to cite this book chapter:

Ancharaz, Vinaye Dey (2025) 'Policy, resources, actors and capacities',
in: Luke, David (ed) *How Africa Eats: Trade, Food Security and Climate Risks*,
London: LSE Press, pp. 67–106. https://doi.org/10.31389/lsepress.hae.d
License: CC-BY-NC 4.0

CAADP has three key emphases:

- Continent-wide coordination of agricultural policies, with support from the AU, that should be market-driven and private sector-led.
- Evidence-based agricultural policymaking underpinned by public investment in infrastructure, research and extension services.
- Modernisation of the farming practices of smallholder farmers, who constitute the bulk of Africa's agricultural producers.

In pursuit of these objectives, policymakers committed to two key targets under CAADP: achieving an average 6 per cent growth in agricultural output per year and allocating 10 per cent of public expenditure at minimum to agriculture. Brüntrup (2011) describes CAADP as 'Africa's attempt to reverse the negative trends in the agriculture sector'. Examples of these negative trends include the sector's sluggish growth and declines in the shares of public spending and official development assistance directed towards the sector.

The CAADP compact included national and regional implementation arrangements. At the country level, AU member states were required to develop national agriculture (and food security) investment programmes (NAIPs). The regional economic communities (RECs) were tasked with including regional agriculture investment programmes (RAIPs) in their activities. In 2014, a CAADP review resulted in the adoption of the AU's Malabo Declaration, which added granular commitments to the original CAADP objectives. These were enhancing finance in agriculture; ending hunger; halving poverty; boosting intra-African trade in agricultural goods and services; enhancing agriculture's resilience to climate variability; and active monitoring of actions and results through biennial reviews. Specific goals and targets, to be achieved by 2025, were set for each commitment.

Regional and international institutions concerned with agricultural development aligned their activities to CAADP and Malabo Declaration. This included the Alliance for a Green Revolution in Africa (AGRA), which was established in 2006 as an inclusive a consortium of stakeholders to build agricultural capacities and provide technical assistance. When the African Development Bank (AfDB) launched its High 5 priorities in 2016, among which was Feed Africa, a strategy for agricultural transformation in Africa for the decade 2016–2025, it broadly aligned with the CAADP's goals and Malabo Declaration commitments. International organisations like the Food and Agriculture Organization of the United Nations (FAO) and the International Fund for Agricultural Development (IFAD) also operate within the CAADP framework.

Assessments of CAADP implementation generally suggest that its impact has been limited (Signé 2017). While most countries had made varying degrees of progress towards the Malabo goals and targets between the first biennial review (BR1) in 2017 and the second (BR2) in 2019, only a handful were on track to meet the goals by 2025 (Makombe and Kurtz 2020). The third

Table 4.1: Progress in achieving the Malabo commitments as assessed in BR3

Commitment area	Number and countries on track	Compared to previous level in BR2
Overall	1: Rwanda	Major deterioration (from 4)
1. Recommitment to the principles and values of the CAADP process	3: Rwanda, Tanzania, Zimbabwe	Slight improvement (from 2)
2. Enhancing investment finance in agriculture	4: Egypt, Eswatini, Seychelles, Zambia	Major improvement (from 0)
3. Ending hunger by 2025	1: Kenya	No change (from 1)
4. Halving poverty through agriculture by 2025	2: Ghana, Morocco	Major deterioration (from 9)
5. Boosting intra-African trade in agricultural commodities and services	4: Botswana, Nigeria, Senegal, Sierra Leone	Major deterioration (from 29)
6. Enhancing resilience to climate variability	15: Burundi, Cabo Verde, Cameroon, Egypt, Ethiopia, Gambia, Ghana, Lesotho, Malawi, Mali, Morocco, Namibia, Rwanda, Seychelles, Zimbabwe	Slight improvement (from 11)
7. Enhancing mutual accountability for actions and results	11: Mali, Ethiopia, Rwanda, Morocco, Mauritania, Tanzania, Tunisia, Senegal, Ghana, Botswana, South Africa	Slight deterioration (from 14)

Source: Author's compilation based on BR3 and BR2.

biennial review (BR3) – in 2021 – reached the same finding. At the time of writing, the result of the fourth bilateral review was not available.

Table 4.1 shows the countries that are on track on each of the seven commitment areas of the Malabo Declaration according to BR3, indicating whether this represents an improvement over BR2. The seven commitments are tracked through 24 targets and 47 indicators. Of the 51 reporting member states, only one (Rwanda) was on track. This is a regression from the four countries that were on track in BR2. While 19 other countries were classified as 'progressive', the continent as a whole was deemed off target.

Member states scored especially poorly in commitment areas 1 to 5. Several countries registered improvements across the three indicators that constitute the first commitment. In general, however, the average score remained below 50 per cent, and the progress achieved was not robust enough to meet the targets set for the 2021 BR. On the second goal, of enhancing investment finance in agriculture, only four countries were on track. While this is an improvement over BR2, where no country was on track, progress on this critical goal is very slow. BR3 confirms that most African countries have fallen short of the CAADP goal of achieving 10 per cent. And the picture appears only to have got worse – according to estimates from FAO, the agriculture budget's share in Africa's total public expenditure was even lower for 2019–2021 than for 2014–2018 (author's cross check with 'SDG Indicators' n.d.). The target for agricultural growth of 6 per cent per annum has also remained elusive (Badiane, Collins and Ulimwengu 2020).

The third goal, the Malabo commitment on ending hunger by 2025, is particularly ambitious in comparison to the UN's Sustainable Development Goal (SDG) 2, on achieving zero hunger globally. The latter allows for a longer time frame of 15 years up to 2030 for ending hunger, compared to just 10 years in the Malabo Declaration. Only one country (Kenya) appears to be on track. The commitment of halving poverty through agricultural development by 2025 may be judged to be equally ambitious. Only two countries reported to be on track, down from nine countries in the BR2 cycle. Conversely, significant progress was noted on the target of achieving at least 6 per cent growth in agricultural value added per year, with 21 countries meeting the target, compared to only three in 2019. This achievement, however, was eclipsed by a major lapse on another indicator, namely the proportion of rural women empowered in agriculture (target 20 per cent by 2025, for which only 10 out of 51 member states that reported on this target were on track in 2020, up from eight in 2018).[1] For the target on creating jobs for 30 per cent of youth in in agricultural value chains by 2025, difficulty collecting data on the indicator meant that only 34 countries reported on the target, of which 17 were on track by 2020 (compared to 13 in 2019 and 14 in 2018). Eleven countries had already achieved the target for 2025 by 2020.

The Malabo goal of boosting intra-African trade in agricultural goods and services is also closely related to the question of regional food security discussed in earlier chapters. In this area, too, there is need for greater progress since only four countries, and none of the regions, were reportedly on track as of 2021. Efforts at improving the conditions for trade are not yet translating into higher volumes of formal regional food trade. While 18 countries were on track to create an enabling environment for intra-African trade, only one country (Nigeria) achieved the target of tripling intra-African trade in agricultural products. The role the African Continental Free Trade Area (AfCFTA) could play in intra-African food trade is discussed in Chapters 5 and 6. The analysis presented suggest that trade liberalisation under the AfCFTA will have only limited impact in boosting intra-African trade. A much stronger impact will be generated by tackling non-tariff barriers through customs, trade facilitation and related border reforms.

The results of BR3 are broadly in line with other empirical assessments of CAADP that are available. For example, an assessment of CAADP implementation against the UN Food System Summit carried out an assessment that focused on five action tracks, namely (1) access to safe and nutritious food; (2) shifting to sustainable consumption patterns; (3) boosting nature-positive production; (4) advancing equitable livelihoods; and (5) building resilience to vulnerabilities, shocks and stress. These action tracks were triangulated with the BR3 performance indicators (Kapuya et al. 2022). The assessment revealed that fewer countries were on track in 2021 than had been in 2019, although the Covid-19 pandemic may have had an impact (Kapuya et al. 2022).

Another study, based on computable general equilibrium modelling, found that the six countries considered – Côte d'Ivoire, Ethiopia, Malawi, Mozambique, Niger and Rwanda – would make only limited progress on the Malabo commitments by 2035. While implementing NAIPs would help, this would not allow all of these countries to meet all CAADP targets. Even Rwanda, the only country found to be on track in BR3, would miss some CAADP targets along with its poverty reduction and equity objectives even if it implemented its NAIP (Diallo and Wouterse 2023). Other assessments reach similar conclusions (Brüntrup 2011; OECD and FAO 2022).

This record engenders scepticism about whether CAADP can live up to its goal of transforming African agriculture. According to Action Aid (2013), to be more effective in supporting agricultural development, African countries could consider programmes targeted to the needs of female and smallholder farmers, as well as exploring the potential of sustainable agriculture, which can carry greater benefits for food security (Adenle, Wedig and Azadi 2019).

However, CAADP has still been useful as a policy initiative. It has provided a comprehensive approach to agricultural development under the auspices of the AU and is mainstreamed into planning at the national and REC levels. The compact provides a basis for mutual accountability. Some development partners including the US are aligning their interventions with CAADP processes. The World Bank operates a dedicated CAADP fund and has stepped up its awareness and capacity-building support (Benin 2018). These aspects are helping CAADP to adapt its implementation experiences and stakeholder expectations. It has been observed that CAADP's foundation on mutual accountability and the framework it provides for aligning external support are probably why the programme remains a rallying policy tool when some other AU initiatives have withered away (Brüntrup 2011).

4.2 Resources

Implementation of the CAADP was initially estimated to require total investment in the region of US$251 billion, or US$17.9 billion per annum, over the period 2002–2015. More recently, AGRA has estimated that Africa would need US$40–77 billion a year in public investment (equivalent to 9–17 per cent of fiscal revenues, or 7–13 per cent of public spending, in Africa as of 2019) and as much as US$180 billion in private investment between 2022 and 2030 to boost agricultural transformation and attain SDGs like ending hunger and halving poverty (AGRA 2022; author's analysis of Economic Commission for Africa, African Development Bank Group and African Union Commission 2021). Reaching these levels of investment calls for ramped-up efforts by governments to mobilise investment and to meet or exceed the CAADP target of 10 per cent of public expenditure allocated to agriculture. It also calls for simultaneous actions on a number of fronts including incentives for the private sector to invest in agriculture and agribusiness, including by reducing risk and subsidising investment, which are allowed under WTO rules on the 'development box' under the Agreement on Agriculture as discussed in Chapter 9; attracting larger flows of foreign direct investment (FDI) in agriculture; and the utilisation of innovative instruments such as risk-sharing, guarantees, quasi-guarantee products like warehouse receipts that enable access to finance, public–private partnership schemes, supply-chain financing, leasing facilities and financial technology (fintech). Resources that are provided through new approaches to financing sustainability in the context of climate change can also be tapped.

Public expenditure

CAADP's 10 per cent public expenditure is an aggregate requirement. It does not distinguish between recurrent spending (such as the cost of providing seeds and fertilisers, research and extension, and training and information), capital spending (such as infrastructure, machinery and equipment) and expenditure on adaptation and mitigation measures related to agriculture in nationally determined contributions (NDCs). In practice, the distinction may not matter. For instance, fertilisers may also be considered a capital expenditure, or investment, since they help restore soil quality and, thus, support enhanced yields in the future (Mengoub 2018). However, it is useful to note that investment tends to have longer-term impacts on productivity. For example, investments in rural infrastructure (roads, transport and storage systems, input supply networks, etc.) are known to support agricultural competitiveness and generate growth.

Figure 4.1 provides an overview of the average share of public expenditure allocated to agriculture during 2017–2021. However, it should be noted that the data does not make a distinction between operational or capital expenditure. Some data is also missing for some years for some countries.

On the whole, the share of public expenditure allocated to agriculture has remained consistently low. Malawi is the only country where the share is above 10 per cent. In Benin, Togo, Central African Republic, Zambia and Guinea-Bissau, the share has averaged above 5 per cent in recent years.

Figure 4.1: Share (in percentage) of public expenditure allocated to agriculture, averages for 2017–2021

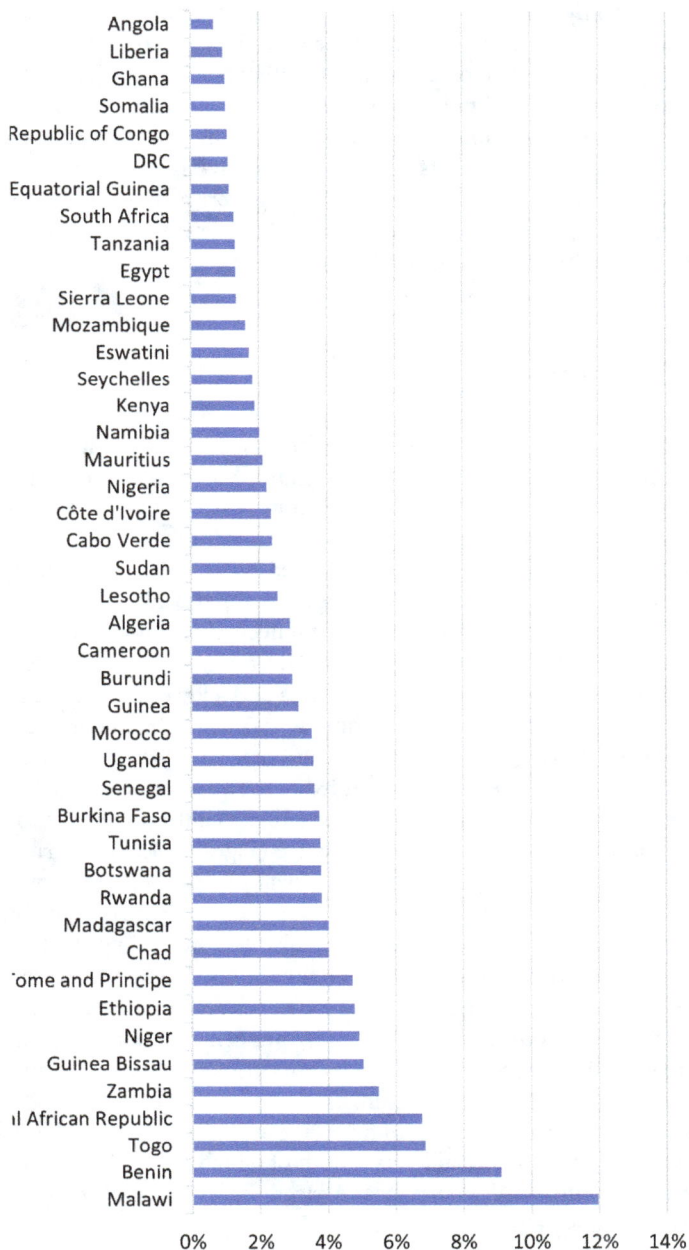

Source: Author's calculations based on FAOSTAT data.
Note: Data for 2020 or 2021 is not available for several countries. In these cases, the average is calculated on data for 2017–2020 or 2016–2019, respectively.

The reasons for the low level of public investment in African agriculture are complex and go beyond the perennial resource constraint and poor policies. Political economy considerations would suggest that the geographically scattered smallholder farmers – the main beneficiaries of agricultural spending – generally lack influence on agricultural policy (Beintema and Stads 2017). Public goods such as technology adoption, market research and rural infrastructure are generally underfunded. Agricultural research, in particular, tends to be neglected despite its high returns on investment in the long run.

Yet, from a political economy perspective, governments also intervene to facilitate food imports to meet food security objectives, as we saw in Chapter 2 and will discuss further in Chapter 8.

Private investment

Reliable data on domestic private investment in agriculture is not available; however, to the extent that the private sector (comprising farms and enterprises at various levels of scale) dominates agricultural investment in many countries, gross fixed capital formation in agriculture (as a share of value added) could serve as a rough approximation for investment by the private sector in agriculture. In Africa, this share has fluctuated between 10 and 12 per cent for much of the past two decades and averaged 10.8 per cent during 2017–2021 (Figure 4.2). The share is low and does not reflect Africa's comparative advantage in agriculture. There is also substantial variation across Africa, with a higher (20 per cent) share in Southern Africa and a lower (6.7 per cent) share in East Africa. Worryingly, since reaching a peak at 12.1 per cent in 2013, the trend has been downward, with the decline worsening sharply since 2019.

At the national level, less than one-third of African countries have agriculture investment shares in value added above 10 per cent, and only 12 countries have a share higher than the African average (Figure 4.3). Southern African countries like Namibia, South Africa, Eswatini, Zambia, Zimbabwe and Mauritius are leaders at the continental level. Some North African countries (including Morocco, Tunisia and Algeria) also feature among the top investors, as do West African countries like Nigeria, Côte d'Ivoire, Cameroon and, to a lesser degree, Senegal and Ghana. Conversely, Eastern African countries rank much lower. Countries like Kenya (5.7 per cent), Ethiopia (6.2 per cent) and Madagascar (5 per cent) boast significant agricultural potential but attract low levels of investment in agriculture.

There are typically two sources of financing available to smallholders – personal savings and commercial loans – both of which are limited. Like micro, small and medium-sized enterprises, African smallholder farms face major barriers to formal credit (Mengoub 2018). Lacking education, knowledge and information, smallholder farmers are typically unable to prepare a viable business plan as a basis for obtaining a bank loan. This makes it difficult for banks to evaluate and price risk appropriately. This adverse combination of factors

Figure 4.2: Share (in percentage) of agricultural investment in value added, 2001–2021

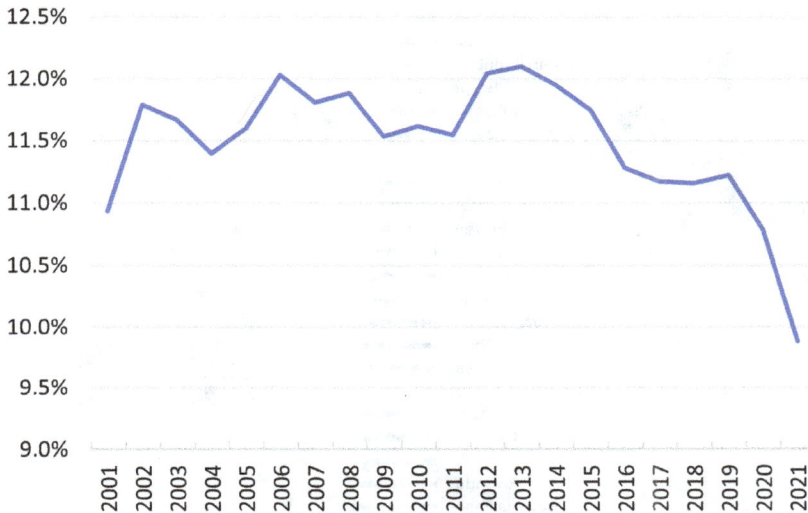

Source: Author's calculations using FAOSTAT data.

along with relatively high levels of inflation generate interest rates as high as 47 per cent, which was the five-year average across Africa for 2017–2021.

Commercial or middle-scale farmers should, in theory, enjoy better access to finance. However, there is a dearth of evidence that mid-scale farmers have better access to finance than smaller farmers even if they are better organised and more educated and try to maximise profits and grow their business (rather than just providing a livelihood for themselves and their families).

The share of bank credit going to agriculture varies widely across countries, with a few countries, notably Malawi, Sudan and Zambia, posting shares averaging 15 per cent or more during 2017–2021. However, at the level of Africa, this share has hovered around 4 per cent, which is a strong indication that very little bank credit flows to the agriculture sector – even in countries that are known to have a strong agricultural vocation (Figure 4.4).

The excessive caution of banks and other financial institutions in providing credit has provided an opening for microfinance institutions (MFIs) and development finance institutions (DFIs) as credit facilitators. The microfinance movement is gaining ground across Africa. Although agriculture may represent a small share of MFIs' portfolios, they nevertheless serve a key role in easing farmers' access to credit. This supports productivity improvement through the acquisition of better-quality seeds, fertilisers and machinery. A two-year randomised controlled trial in Chipata, Zambia, suggests that farming households that had access to microcredit produced on average 8 per cent

Figure 4.3: Share (in percentage) of agricultural investment in agricultural value added, averages for 2017–2021

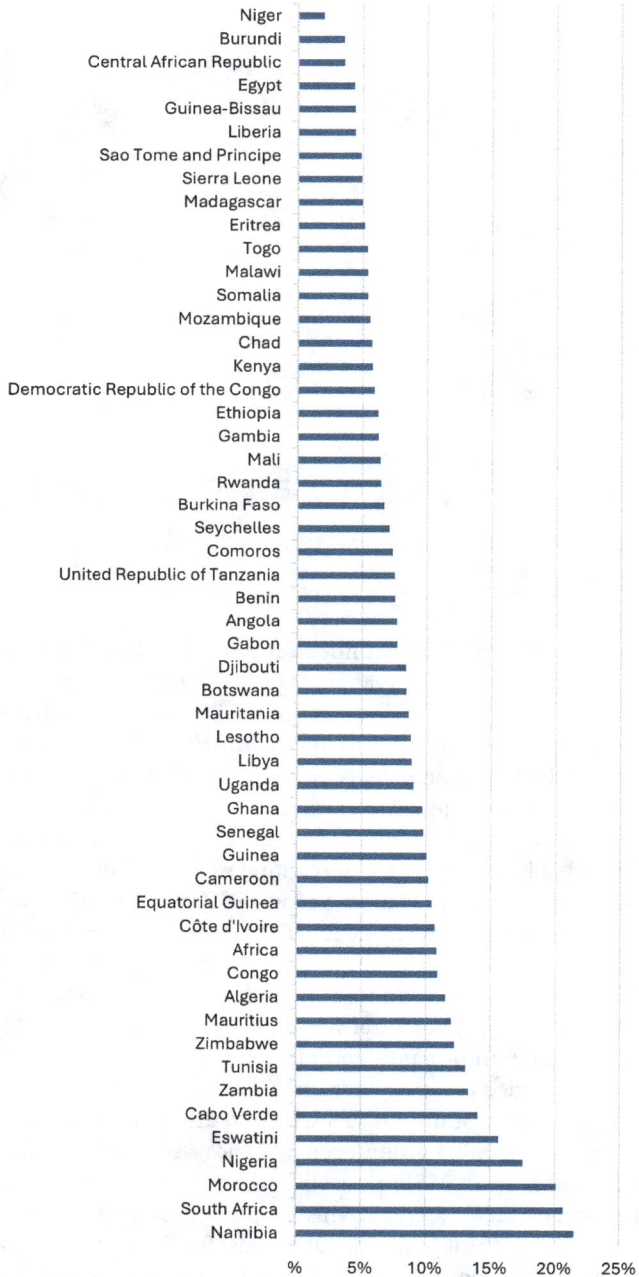

Niger
Burundi
Central African Republic
Egypt
Guinea-Bissau
Liberia
Sao Tome and Principe
Sierra Leone
Madagascar
Eritrea
Togo
Malawi
Somalia
Mozambique
Chad
Kenya
Democratic Republic of the Congo
Ethiopia
Gambia
Mali
Rwanda
Burkina Faso
Seychelles
Comoros
United Republic of Tanzania
Benin
Angola
Gabon
Djibouti
Botswana
Mauritania
Lesotho
Libya
Uganda
Ghana
Senegal
Guinea
Cameroon
Equatorial Guinea
Côte d'Ivoire
Africa
Congo
Algeria
Mauritius
Zimbabwe
Tunisia
Zambia
Cabo Verde
Eswatini
Nigeria
Morocco
South Africa
Namibia

%   5%   10%   15%   20%   25%

Source: Author's calculations based on FAOSTAT data.

Figure 4.4: Share (in percentage) of agriculture in total bank credit in Africa

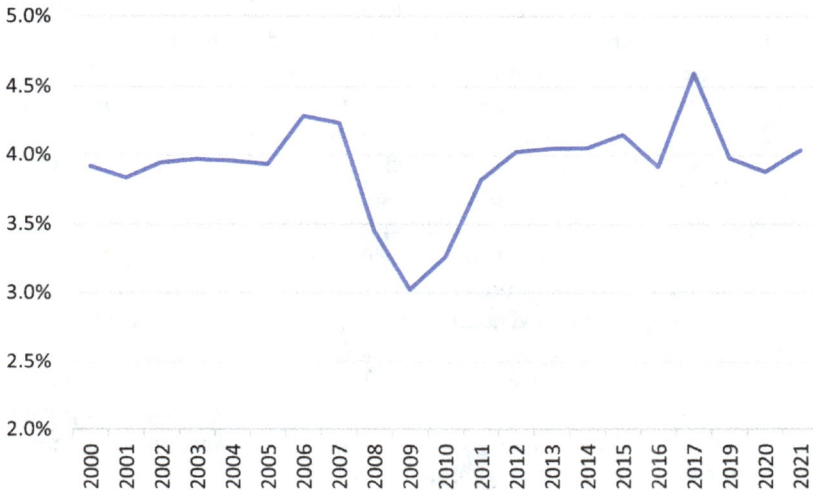

Source: Author's calculations using FAOSTAT data.

more than those in villages without such access (Stewart 2020). MFIs have also had a transformative impact on women in agriculture, empowering them with financial resources and training, and breaking gender barriers. In Uganda, Kenya and Tanzania, for example, the Women's Microfinance Initiative has helped aspiring women entrepreneurs in the food supply chain to build an income-generating business to improve household living standards. The initiative has resulted in a fivefold increase in clients' incomes in some cases within a relatively short period of time (World Bank 2018). However, other studies have found more ambiguous effects of microfinance programmes in Africa's agricultural sector (Economic Commission for Africa 2019, pp.29–32; van Rooyen, Stewart and de Wet 2012). And some MFIs have sparked controversy for charging excessive interest rates or demanding collateral that borrowers are incapable of providing. Banerjee et al. (2015) argue that microfinance borrowers are likely to be subsistence or 'reluctant' entrepreneurs rather than 'gung-ho' or transformational ones, which limits the impact of microcredit on entrepreneurship and poverty alleviation.

Moreover, instead of channelling scarce development finance towards micro-enterprises that are not always very productive, governments may wish to focus on supporting high-potential businesses that have a good chance of raising living standards on a much broader scale. These can capture export market share, reduce the cost of food for domestic consumers and pay decent wages based on high worker productivity (Economic Commission for Africa 2017; Economic Commission for Africa 2019, pp.29–32).

With a mandate to de-risk investments, DFIs are playing a critical role in deepening financial services for Africa's farming community, especially smallholder farmers. Some DFIs have attracted significant amounts of donor funding while others have departed from their mandate and taken an increasingly commercial route as funding from public sources thinned out. However, DFIs face challenges of their own, which significantly limit the support they could provide to smallholders. Most of them are urban-based and thereby removed from their agricultural constituents. While frequent field visits by liaison officers can resolve this problem, DFIs do not invest sufficiently in this cadre of personnel, or in specialised investment professionals, who can help develop a pipeline of bankable projects. There is scope for DFIs to adopt more innovative financial products that are tailored to the unique needs of smallholder farmers, and leverage partnerships with donors and community-based organisations working closely with farmers (Savoy 2022).

There is emerging evidence of financial flows into African agriculture from a variety of nontraditional sources. These include venture capital and private equity funds, innovative instruments such as value chain financing, green bonds, insurance and credit guarantee schemes, blended finance, impact investment funds, and fintech solutions such as crowdfunding, peer-to-peer lending, and mobile payment applications. A recent report reveals that venture capital investment flows doubled in 2021, albeit from a low base in 2020 (AgFunder 2022). Although, in absolute terms, the amount represented less than 1 per cent of global venture capital spending on agriculture, it is nevertheless encouraging since investment of this type was negligible just a decade ago and there are signs that it is growing (Grow Further 2022). Private equity investment in agriculture is also gaining prominence across the continent, with the rise of equity funds, such as the African Agricultural Capital Fund, and private equity firms like Phatisa and Sahel Capital. These firms have demonstrated success in supporting agribusiness enterprises, emphasising sustainable and impactful investments (Phatisa 2021).

Innovative financing instruments play a crucial role in addressing the diverse needs of the agricultural sector. Value-chain financing, for instance, involves providing financial services to actors along the agricultural supply chains such as farmers, processors, and distributors (SME Finance Forum 2017). Value chain financing can take into account existing relationships in the value chain to reduce the perceived risk of the investment (Cuevas and Pagura 2016, p.50). A good example is the Partnership for Inclusive Agricultural Transformation in Africa, which utilises value chain financing to enhance financial inclusion in the farming sector.

Green bonds have emerged as sustainable financing options for African agriculture, aligning with the sector's growing emphasis on environmental responsibility and climate-smart practices. Impact investment funds are blended finance initiatives that combine public and private funds to achieve a financial return along with targeted social or environmental impacts. They focus on projects that contribute to sustainable development, poverty alleviation, and environmental conservation. The FAO (2018) notes that agricultural investment funds have flourished around the world, including in Africa, aided by investors' searching for impact opportunities. Several case studies have documented the developmental impacts of these funds, which are 'fast becoming the vehicle of choice for governments and donors looking to invest in African agriculture and encourage private sector investors to do the same' (Castell 2019).

Finally, fintech can potentially revolutionise agrifinance by introducing digital solutions to traditional challenges. M-Pesa (a mobile money service in seven African countries), for instance, has provided smallholder farmers with a convenient and secure means of paying and receiving cash in regions where access to banking services is limited. However, while numerous studies have documented the positive impacts of M-Pesa, including on poverty and rural women's empowerment, empirical evidence of the use of mobile financial services for agricultural activities has been scant. A rare, recent study based on nationally representative data from Kenya reveals that, while more than 80 per cent of Kenyan farmers use mobile money, less than 15 per cent of them use it for agriculture-related payments. Moreover, mobile loans for agricultural investment are used by less than 1 per cent of farmers (Parlasca, Johnen and Qaim 2022). This suggests that the use of mobile financial services in agriculture is lower than commonly perceived and a transformative impact on smallholder farming is yet to emerge. Similarly, innovative financing models, such as crowdfunding and peer-to-peer lending platforms, have opened new possibilities for agricultural start-ups in Africa, but their potential remains to be harnessed.

Foreign direct investment (FDI)

Data on FDI in Africa's agriculture sector is patchy. It nevertheless shows that agricultural FDI as a share of total FDI inflows is as low as 0.025 per cent for Nigeria to 3.9 per cent for Tanzania.[2] At the continent-wide level, in 2022, less than 2 per cent of FDI to new subsidiaries ('greenfield' investments) in Africa, and around 3 per cent of incoming international project finance flows into the continent, went to agri-food systems (United Nations Conference on Trade and Development 2023). It seems that the appeal of the extractive sector in some countries has proved a bane for agriculture. Elsewhere, fiscal incentives to attract FDI into manufacturing or services have had the effect of crowding out the agriculture sector. In Mauritius, for example, incentive schemes to attract FDI into property development since 2004 have been overly successful such that the country receives hardly any FDI in the productive sectors. According to FAO data, only 0.27 per cent of FDI inflows to Mauritius between 2017 and 2020 went to agriculture, forestry and fishing.

According to UNCTAD Stat, aggregate FDI inflows to Africa in 2021 represented a mere 5.2 per cent of global FDI flows. At the regional level, East Africa received the lowest share of aggregate FDI inflows to Africa (an average 15 per cent during 2017–2021). While FDI was fairly evenly distributed among the other four regions of Africa, there is strong evidence of concentration in South Africa, which accounts for 96 per cent of FDI inflows to Southern Africa. In Central Africa, Congo, Democratic Republic of the Congo and Gabon received 77 per cent of the region's FDI inflows. At the continental level, five countries (South Africa, Egypt, Congo, Ethiopia and Ghana) accounted for 56 per cent of all FDI flows during 2017–2021. With the exception of Ethiopia, a common feature of these top FDI destinations is that they are all major commodity-producing countries, with FDI targeted at the minerals sector rather than at agriculture and food production. This suggests that much of African FDI is resource-seeking, with the extractive sector as the magnet (Gerlach and Liu 2010). However, as of 2022, the majority of 'green-field' FDI and international project finance deals directed to Africa went to the continent's energy sector (i.e. producing energy for use on the continent, not extracting fossil fuels form the ground) (United Nations Conference on Trade and Development 2023).

There is a dearth of empirical evidence on the impact of FDI on agriculture and food security. A review of case studies paints a mixed picture, with the impacts varying significantly across countries, depending on the terms of the investment, the type of business model and the institutional framework in place in the host country. A case study of eight countries – Egypt, Ghana, Madagascar, Mali, Morocco, Senegal, Sudan and Uganda – provides some evidence that FDI in agriculture generated benefits such as employment creation, higher productivity, improved access to finance and markets for smallholders, and technology transfer. However, these impacts varied across countries and across locations within a given country (Gerlach and Liu 2010). For instance, job creation was correlated with the capital intensity of invest-ment projects, but FDI in Mali substituted local labour for foreign (Chinese) workers, while farmers displaced by land acquisitions in Madagascar were unable to find other employment.

Husmann and Kubik (2019) finds that agricultural FDI flows to Africa tend to be positively correlated with the size of the domestic market, the contracted plot size, and the quality of infrastructure and institutions, and have posi-tive impacts on farm and labour income and on technical innovation. There is much less evidence on the impact of agricultural FDI on food security. A rare study based on panel data from 56 developing countries (not all Afri-can) found that FDI in agriculture has a mixed effect on food security in the host country, but that the impact is more favourable where land governance systems are well established (Dogan 2022). The findings suggest that land tenure reforms that formalise customary land rights, and mechanisms that ensure greater transparency of agricultural investment processes, can enhance the impact of FDI on food security.

Some investment deals that require land acquisition lacked transparency. That is, they were not accompanied by appropriate impact assessments, resulting in smallholders being displaced or dispossessed of their land or in other adverse impacts on local communities and the rural environment. However, there is evidence to suggest that the so-called 'land grabs' in Africa have not provided the returns that were expected. The continent features the highest proportion of 'failed' land deals. These are investment contracts or negotiations that were cancelled, partly because of disputes with local communities (Feyertag and Bowie 2021). In fact, half of all 'failed' transnational agricultural land deals between 2000 and 2020 occurred in sub-Saharan Africa (Lay et al. 2021).

Several factors have contributed to the low level of agricultural investment in Africa including weak land laws and governance institutions (Maina 2022; WEF 2016).

Foreign aid to African agriculture and food security

Official development assistance (ODA) aid disbursement to the agricultural sector in Africa is small and has never surpassed 8 per cent of total aid flows to Africa since 2001 (Figure 4.5). After edging steadily up from a low

Figure 4.5: Aid disbursement to the agriculture sector in Africa as a share (in percentage) of total disbursements, 2000–2021

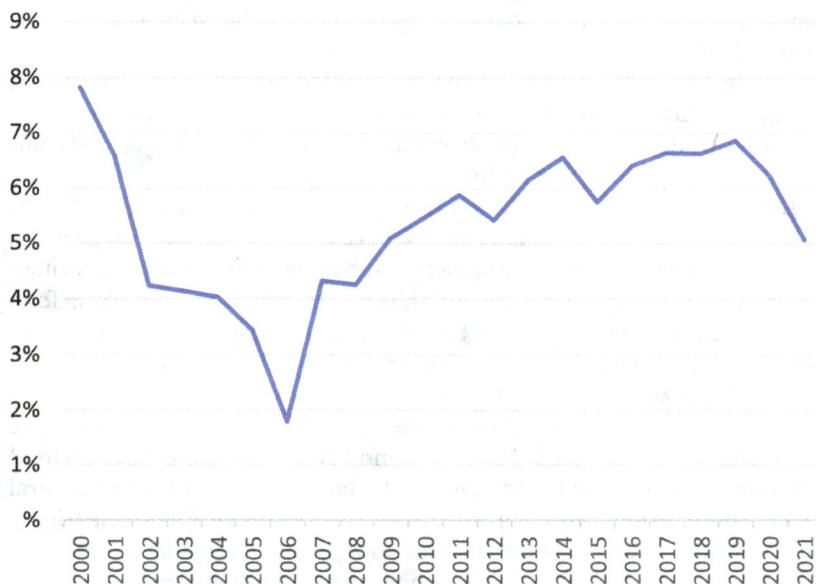

Source: Author's calculations based on FAOSTAT data.

Figure 4.6: Aid flows to agriculture: Commitment vs. disbursement (US$ billion, current prices), 2000–2021

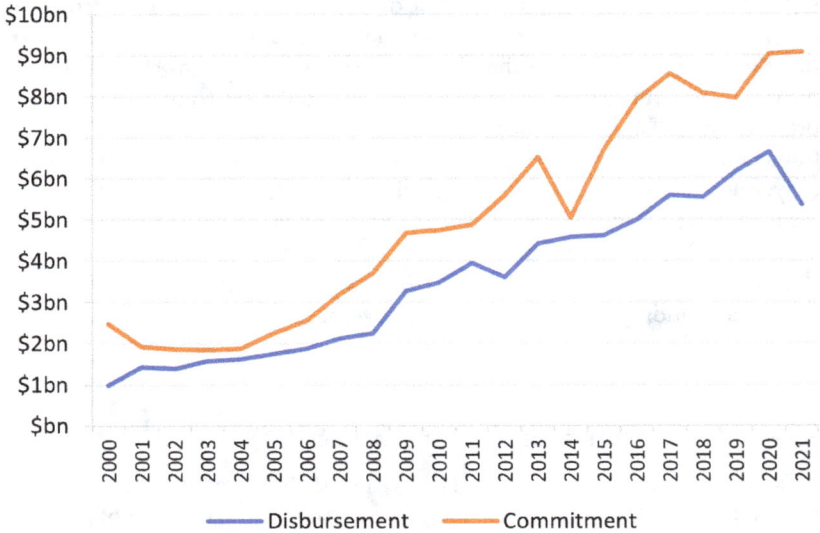

Source: Author's calculations based on FAOSTAT data.

of 1.8 per cent in 2006, the share has fallen since 2019 to reach 5.1 per cent in 2021. Figure 4.6 shows that, while aid commitments and disbursements have generally increased over the years, the gap between the two has also widened.

During 2017–2021, disbursements amounted to two-thirds of commitment levels, revealing an important gap between what donors promise and what they actually deliver. African governments should call for scaled-up and additional aid to support the agriculture sector.

There are two distinct pathways through which foreign aid (or ODA) is channelled to African agriculture and food security. The first pathway provides resources for agricultural development and building food security capacities. The second pathway is food aid, which is specifically aimed at making food available including through cash transfers such as balance of payments support for financing food imports. This section is on the first pathway, foreign aid to African agriculture; the next section is on food aid.

The empirical literature supports the view that development assistance to agriculture is beneficial. McArthur and Sachs (2018) use three stylised scenarios to show how foreign aid can be targeted to support agricultural productivity through optimal input use. They find that ODA can trigger an expansion of the agricultural sector and generate permanent productivity and welfare gains, which could render such aid unnecessary in the long run. In the same vein, an econometric study based on 47 African countries finds similar effects of agricultural aid on GDP and productivity. There is also some evidence that bilateral aid has bigger productivity effects than multilateral aid (Alabi 2014). Some of Africa's bilateral partners in food and agricultural trade have technical cooperation schemes in place alongside the mutual trade transactions, as discussed in Chapter 8. Ssozi, Asongu and Amavilah (2019) suggest that better host institutions and liberalised markets are prerequisites for ensuring that the impact of aid on agricultural growth and food security is maximised.

At least four distinct aspects of foreign aid to African agriculture can be identified. The first is Aid for Trade, an initiative sponsored by the WTO and monitored by the Organisation for Economic Co-operation and Development (OECD).[3] A significant share of Aid for Trade from both bilateral and multilateral sources goes to rural infrastructure such as roads, irrigation systems and storage facilities. For example, the AfDB's 'Feed Africa' strategy has committed to investing US$24 billion in the next decade to agricultural infrastructure to boost agricultural productivity, reduce post-harvest losses, and support marketing processes.[4] Although included in OECD–WTO reporting as Aid for Trade, these investments also benefit production for local markets.

Second, ODA can support mitigation and adaptation in the agricultural sector, enabling farmers to adopt modern best practices. Box 4.1 provides a summary of the emerging role of climate finance, which has profound implications for agricultural development. Climate-related initiatives through specialised bilateral and multilateral sources are increasingly being directed to support agricultural development. For example, AGRA is working with host governments and local non-governmental organisations (NGOs) to promote the use of improved seeds, appropriate fertilisers and modern farming techniques across the continent. The International Centre for Tropical Agriculture collaborates with national agricultural research agencies to develop climate-smart solutions such as the implementation of drought-resistant crop varieties and soil conservation techniques, enabling farmers to adapt to climate change. The World Agroforestry Centre is assisting African farmers through its Evergreen Agriculture initiative – an approach that combines tree planting with agricultural support – to increase crop yields, improve soil fertility and diversify income sources.[5]

Third, ODA can facilitate access to finance and credit for smallholder farmers. For example, the IFAD provides financial support and technical assistance to small-scale farmers in Africa, enabling them to improve their farming techniques, diversify income sources and, ultimately, achieve food security for their households and communities.[6]

Fourth, foreign aid can empower women in agriculture. Several bilateral development partners specifically require that their resources benefit women. The UN's Women's Entrepreneurship Development Programme is an example of an initiative that is aimed at supporting women farmers in Africa through training, financial services and access to markets.[7]

Box 4.1: Climate finance

It is estimated that US$2.8 trillion will be needed from 2020 to 2030 to implement the commitments African countries have made in NDCs as part of the implementation of the Paris Agreement. Mitigation accounts for two-thirds of reported climate finance needs for the period 2020–2030, distributed across the following four sectors: transport (58 per cent), energy (24 per cent), industry (7 per cent) and agriculture, forestry and other land use (AFOLU, 9 per cent). Adaptation represents only 24 per cent of total climate finance, even though for Africa adaptation, rather than mitigation, remains the dominant priority. African governments have committed to contributing 10 per cent of the total cost of climate action. This means that US$2.5 trillion (or an average of US$250 billion annually) needs to be mobilised externally from climate funds and donor support. In 2020, Africa's climate finance flows, both domestic and international, totalled US$30 billion, or about 12 per cent of the need. The funding gap is significant.

Climate finance remains central to addressing climate change equitably and efficiently, including achieving adaptation goals. Multilateral climate funds, such as the Least Developed Countries (LDC) Fund, the Special Climate Change Fund (SCCF), the Adaptation Fund and the Green Climate Fund (GCF) are the main global initiatives dedicated to combating climate change. Several developed countries have also launched climate finance initiatives of their own or are providing climate finance through bilateral development assistance agencies. Examples include the International Climate Fund (United Kingdom), the Hatoyama Initiative (Japan) and the Global Climate Change Alliance (European Commission) (Watson and Schalatek 2020). Complementing these channels of climate finance are non-concessional lending by multilateral development banks; bilateral non-concessional lending (government-to-government loans); and international private finance (e.g. equity investments or external loans) (Ahluwalia and Patel 2022, p.317).

At the latest (at the time of writing) Conference of the Parties (COP) 28 in November 2023, new climate finance commitments were made. The second replenishment of the GCF was boosted by new pledges, taking total commitments to US$12.8 billion. New commitments to the LDC Fund and the SCCF amounted to US$174 million, while the Adaptation Fund attracted US$188 million in pledges. The main COP 28 highlight was the agreement on the operationalisation of a loss and damage fund that had until then proved elusive. By the time of writing, it had received pledges of up to US$700 million, an amount too small when compared to the projected economic costs of loss and damage, which should be between US$290 billion and US$580 billion for developing countries, according to one set of estimates (Markandya and González-Eguino 2019; UNFCCC 2023).

(continued)

(continued)

The multilateral climate funds have been criticised for their lack of transparency (Transparency International 2022) and limited consultation with the civil society and indigenous communities (Kumar 2015). For example, the GCF does not have a disclosure policy or accountability mechanism. The World Bank, which will house the Loss and Damage Fund, has also been criticised for failing to account for 40 per cent of its reported climate spending (Harvey 2022). Some critics have noted that a substantial amount of climate finance flows through international institutions and multilateral banks instead of being sent directly to the project implementers on the ground. (However, previous research suggests that there is no clear evidence that bilateral aid is better than multilateral aid, or vice versa, and that multilateral aid earmarked for a specific purpose may achieve the best of both worlds (Biscaye, Reynolds and Anderson 2017; Gulrajani 2016).)

Food aid

Historically, food aid emerged as a response to acute emergencies, such as conflicts, natural disasters and famines, with the primary objective of saving lives and preventing starvation (WFP 2021a). Over time, food aid efforts have evolved to incorporate a developmental dimension aimed at enhancing long-term food security and fostering sustainable agricultural practices (Barrett and Maxwell 2007).

Food aid can be categorised according to its intended objectives and supply methods. In relation to objectives, three aspects can be identified. First, programmatic food aid is provided for balance of payments or budgetary support to finance food imports. The second is project food aid, which targets poverty alleviation and disaster prevention for specific vulnerable groups or areas. And the third is relief aid, provided for distribution to disaster victims. These distinctions can be blurred, especially in crises. In relation to supply methods, food aid includes direct transfers from donors, exchanges between countries, and local purchases for domestic distribution (Organisation for Economic Co-operation and Development n.d.).

Given Africa's vulnerability to food insecurity, the region is a major recipient of food aid, accounting for almost two-thirds (63.3 per cent) of all food aid provided to developing countries between 2017 and 2021. After declining during the pandemic years, food aid has bounced back, reaching US$1 billion in 2022 (Figure 4.7). Eastern Africa attracted 60 per cent of all food aid to Africa in recent years, of which a quarter went to Ethiopia alone.

The effect of food aid on food security is mixed. On one hand, food aid plays a critical role in mitigating acute food shortages and preventing immediate hunger-related fatalities (FAO 2021). For instance, during the 2011 Horn of

Figure 4.7: Development food assistance to sub-Saharan Africa (SSA)

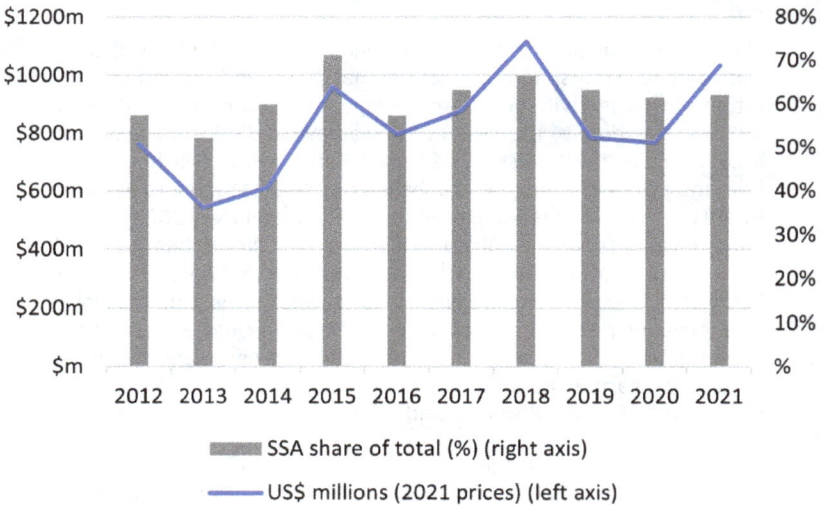

SSA share of total (%) (right axis)

US$ millions (2021 prices) (left axis)

Source: Author's calculations based on data from the OECD Creditor Reporting System database.

Africa drought, food aid helped avert a major humanitarian catastrophe by providing essential sustenance to vulnerable populations (Béné, Devereux and Sabates-Wheeler 2012). During the 2014 Ebola outbreak in West Africa, food aid contributed significantly to the containment of the epidemic by ensuring that affected communities received adequate nutrition (FAO 2014). Food aid can also provide a safety net and bring about unexpected positive outcomes. In Malawi, for example, a school feeding programme to combat malnutrition among children has led to improved school attendance (WFP 2021b). As noted in Chapter 1, the WFP was awarded the 2020 Nobel Peace Prize for its efforts to combat hunger during the Covid-19 pandemic and more generally in conflict-affected areas.

However, the overall effect of food aid on food security is subject to debate. Some critics have argued that food aid, if not properly managed, can undermine local agricultural production by flooding markets with imported goods, which in turn may depress prices and disincentivise local farmers (Barrett and Maxwell 2007). This phenomenon is well illustrated by the case of Malawi, where the influx of food aid disrupted local markets and discouraged farmers from investing in crop production (Jere 2007). In Burkina Faso, the arrival of food aid caused a decline in cereal prices, with adverse impacts on producers and traders (Béné, Devereux and Sabates-Wheeler 2012). The global implications of subsidised food production are discussed in Chapter 9 on the WTO legal framework and food security.

To move beyond the short-term relief offered by food aid and achieve sustainable food security in Africa, more comprehensive and holistic strategies are needed. Targeted support for both agricultural production and social safety nets are strategies that are increasingly being applied (Mogues, Fan and Benin 2015). For example, the Purchase for Progress initiative by the WFP encourages the procurement of food from local sources, thus boosting agricultural production and the local economy. Moreover, the OECD (2006) estimates that food aid in kind entails efficiency costs in excess of 30 per cent; thus, switching to local sourcing of food, where possible, can generate substantial efficiency gains. Nevertheless, the best approach to procuring food aid (local vs. regional vs. long distance) can depend on the context and the programme's objectives (Harou et al. 2013; Lentz, Passarelli and Barrett 2013).

4.3 Actors and capacities

Capacities play a key role in the functioning of food systems that underpin food security. It will be recalled that food systems were defined in the Preface as the sum of actors and interactions along the food value chain – from input supply and production of crops, livestock, fish and other agricultural commodities to marketing, transportation, processing, wholesaling, retailing, preparation of foods, consumption and disposal (AGRA 2022). Several institutions with varying capacities, challenges and opportunities are among these actors. These include farmers operating as smallholders or at a larger scale, functioning as contract farmers or organised in cooperatives. Actors also include market intermediaries, commodity exchanges, marketing boards and agribusiness multinationals that mediate markets and trade. This section reviews the role of these actors in how Africa eats.

Smallholder farmers

Smallholder farmers, operating on family land plots of less than five hectares, provide the foundation for African agriculture and food security. They produce up to 90 per cent of the continent's food and therefore play a crucial role in how Africa eats (IAASTD 2009). Yet many exist in a perpetual cycle of poverty. Over 80 per cent of smallholder farmers produce at the subsistence level (Oyewole 2022). Lacking skills and resources, smallholders are often unable to take advantage of agribusiness opportunities or fully commercialise their output, thereby producing well below their potential (Malhotra and Vos 2021). Inadequate supporting policies and weak institutions remain overarching barriers to the transformation of African food systems (Ulimwengu, Nwafor and Nhlengethwa 2022). This is one of the main reasons why the 'green revolution' has largely bypassed Africa.

But this is not to suggest that smallholder farmers are unproductive. There is evidence to suggest that small farms can be more productive depending on the context and level of technological development (Larson et al. 2014; Fan and Rue 2020). Indeed, African smallholders encounter significant challenges along the agricultural value chain, at both pre- and post-production stages.

Pre-production, small plot sizes preclude economies of scale and make investment in equipment and irrigation unviable. Lack of access to credit, limited technical knowledge about inputs and poor information about input prices are major limitations farmers face in purchasing and using appropriate inputs in the right quantity. Post-production, smallholders often fail to obtain a fair value for their produce and remain vulnerable to downstream actors and high rent extraction. Other challenges include deficient storage facilities, resulting in post-harvest losses averaging 30 per cent of production, according to some estimates (Oyewole 2022). Lack of information on markets, weak linkages to regional markets and product quality are other difficulties (de Brauw and Bulte 2021). These privations generate the conditions for informal markets to thrive. These markets are a ubiquitous feature of African food markets including for cross-border trade as discussed in Chapter 5.

Medium-scale farmers

Medium-scale farmers are very often agricultural entrepreneurs who engage in farming as a business. Their rise has been triggered by the opportunity created by a surge in food prices (Muyanga and Jayne 2018) and the emergence of a mainly urban-based entrepreneurial class (Jayne et al. 2016). Survey evidence from Zambia and Nigeria suggests that medium-scale farmers have plot sizes greater than 10 hectares (Goedde, Ooko-Ombaka and Pais 2019; Jayne et al. 2014). As better-informed entrepreneurs, medium-scale farmers have better access to inputs, technology and markets. In Tanzania and Zambia, medium-scale farmers account for about 40 per cent of agricultural output (Jayne et al. 2016). There is evidence to suggest that the activities of medium-scale farmers have a positive impact on the rural economy mainly through local sourcing for labour, services and other inputs. But there is also evidence, notably from Ghana, that the rise of medium-scale farmers displaces smallholders (Hall, Scoones at Tsikata 2017).

Contract farmers

Contract farming describes a situation where farmers sign a contract with a purchaser, under which the farmer 'commits to producing a given product in a given manner and the buyer commits to purchasing it' (ActionAid 2015, p.3). Compared to smallholder or medium-scale farming, where the farmer takes all the risks associated with the production and marketing, under contract farming these risks are substantially transferred to the buyer (Meemken and Bellemare 2019). The farmer may also benefit from technical assistance, inputs and credit provided by the buyer (Minot 2015). Contract farming is essentially based on the out-grower model.

Contract farming has been hailed as a 'win–win' business model (Hall, Scoones and Tsikata 2017). It provides a ready market for the farmer's produce at a guaranteed price while ensuring a means of secure supply to the buyer for processing and other downstream activities.

Proponents of contract farming argue that engaging famers at different levels of scale, from smallholders to medium-scale farmers, provides them with an environment conducive to productivity improvement, growth and diversification into high-value commodities. Critics argue that the balance of power in out-grower schemes is often harmful to smallholders. Large agribusiness companies wield substantial monopsony power that allows them to force lower prices onto the farmers than they would receive in more open markets. In some cases, smallholders may be excluded from contract farming, leading to their marginalisation and causing income inequality in rural areas (Minot 2015).

A study of cassava growers in Ghana showed that contracts that simply guarantee a market for smallholders' output are not sufficient to ensure mutually beneficial outcomes. Inclusive contracts provide welfare benefits and embed targeted technical services (Poku, Birner and Gupta 2018). But this finding is at variance with results from a field experiment on contract farming in the rice sector in Benin, which show that even the simplest contract has important impacts since it eliminates commodity price risk, giving farmers comfort and confidence to address other constraints on their own (Arouna, Michler and Lokossou 2021).

On the whole, the evidence indicates that contract farming comes with challenges on both sides of the contract: high rates of turnover of schemes, legal restrictions on direct contact between farmers and their contractors, side-selling by smallholders in violation of their contracts, risk of default on the part of buyers when market prices fall below the contracted price, difficulty of dealing with geographically dispersed farmers (Minot 2015). However, if contract farming is tailored to the local context and is inclusive in its reach, it can be an important contribution towards enabling smallholders to increase, diversify and market their production and for fostering agribusiness development in Africa.

Farmer organisations or cooperatives

By joining forces, farmers can exercise leverage in input and output markets. While the cooperative movement has strong roots in many African countries going back to the colonial era, its history has been chequered. After independence, some cooperatives became instruments of political patronage, which, along with food price controls, undermined their effectiveness. Structural adjustment reforms of the 1980s and 1990s ridded cooperatives of failed policies, but revenue loss and falls in membership and viability persisted through the mid-2000s (FAO 2010). Recent years have seen a revival of cooperatives alongside the policy framework provided by the CAADP initiative (Mercier 2020).

Agricultural cooperatives vary in form and functionality. Most are focused on production, including the purchase and sharing of agricultural inputs and equipment, or marketing, which has been their traditional role, or both. Saving and credit cooperatives (SACCOs) are also active in rural communities. While these may not conform to the traditional understanding of the role of farmers' cooperatives, they are a vital support to agriculture, providing farmers with much-needed funding, both for investment and to sustain household consumption during the growing season. SACCOs have witnessed rapid growth in many countries and are becoming the largest part of the cooperative sector (Mercier 2020).

The evidence on the impact of cooperatives is mixed. On the one hand, some studies suggest that cooperatives boosted farmers' bargaining power, empowering them to attract institutional buyers for their products (World Bank 2007). UN organisations such as IFAD and NGOs working in agriculture report that cooperatives have helped smallholders reduce costs and reach larger markets, improving their incomes and food security. In many countries, they are used as a conduit by government and NGOs for farmers' training, knowledge transfer, and research and extension services including those directed at women and youth (Sifa 2014; UN Women 2020). Evidence from South Africa suggests that NGO-supported cooperatives have fared better than those controlled by the government (Sikwela, Fuyane and Mushunje 2016). In Eastern and Southern Africa, cereal marketing cooperatives are seen as more effective at inducing commercialisation than macroeconomic and trade policy interventions (Barrett and Mutambatsere 2008).

On the other hand, a study of the impact of marketing cooperatives on smallholder commercialisation of cereals in rural Ethiopia found that cooperatives secured higher prices but did not achieve any significant increase in the share of cereal production. It is suggested that farmers *reduced* their marketed output in response to higher prices (Bernard and Taffesse 2012).

Market intermediaries

Market intermediaries or 'middlemen' play an important role in agricultural marketing in many parts of Africa. They link farmers to traders and final markets, providing valuable feedback to farmers in addition to critical facilities such as warehousing, insurance and finance. However, intermediaries have often been described as opportunistic agents who profit at farmers' expense and drive commodity prices up.

This perception is often the result of the inefficiencies in African agricultural market systems, and may be erroneous (Eleta 2020). Examining the view that intermediaries exploit farmers by exercising monopsony power, Enete (2009) finds that cassava farmers in a sample of African countries typically sold more through intermediaries than in their absence. Moreover, cassava prices were found to be more stable in Nigeria, where intermediaries competed for farmers' produce, than in other countries where the 'middleman' culture was lacking. Abebe, Bijman and Royer (2016) provide corroborating evidence from Ethiopia. Although they find that gross profit for farmers was, on average, 225 per cent higher without intermediation, they attribute this outcome to better-quality inputs and better contractual arrangements, suggesting that the more well-endowed farmers self-selected into trading directly with wholesalers.

The Economist (2022) goes in the same direction, describing intermediaries as the 'invisible links' in African agriculture and the 'human infrastructure' of African economies. Anecdotal evidence from Ugandan coffee farmers suggests that they nurture a relationship of trust with 'middlemen', who often assume the role of non-existent agricultural banks, providing cash when it is needed.

Some critics have called for a 'better class of middlemen' or cutting them out altogether (Cordaid 2021). Mitchell (2019) argues that 'middlemen', as part of an ecosystem of 'inclusive intermediaries', can play a key role in the commercialisation and industrialisation of agriculture. But this needs to take the form of multi-stakeholder partnerships, involving government and non-state actors.

Commodity exchanges

Commodity exchanges are organised markets where future delivery contracts for specific agricultural products are bought and sold. They range from simple auctions, providing a platform for small farmers to sell their produce at quasi-market-determined wholesale prices, to more sophisticated derivatives markets, allowing participants to hedge commodity price risk. Commodity exchanges act as coordination mechanisms, enhancing information flow, reducing transaction costs and smoothing short-term price variability. They also enhance liquidity by allowing trade in futures contracts (Rashid, Winter-Nelson and Garcia 2010).

Africa was home to the world's first commodity exchange – in Alexandria, Egypt, more than 150 years ago. However, it was not until the post-structural adjustment era that a renewed focus on liberalised markets brought them back into the limelight, with a first wave of 'modern' commodity exchanges taking hold in Zambia, Zimbabwe and South Africa in the early 1990s. Ethiopia established a commodity exchange in 2008 in what may be described as the second wave (Rashid, Winter-Nelson and Garcia 2010). A third wave may be underway as new national commodity exchanges are being developed in Ghana, Tanzania, Nigeria, Kenya and Malawi alongside subregional (e.g. the East Africa Exchange) and continental (the Agricultural Commodity Exchange for Africa) initiatives (Songwe 2011). The latter is a proposal for a network of commodity exchanges complete with warehouse receipt systems functioning across major commodity-producing countries in Africa.[8] These exchanges could be merged into a single platform to create a virtual continental network.

An assessment of commodity exchanges in Ethiopia, Kenya, Malawi, Uganda and Zambia reveals that all five have 'drifted far from the original model' and, except for Ethiopia, have fallen short of their objectives (Robbins and Catholic Relief Services 2011). They have neither improved farmer linkages to formal markets nor generated new opportunities or trading relationships, nor substantively increased farmers' incomes. Some of the commodity exchanges did not develop beyond a platform for disseminating market information; others were not linked to a viable warehouse receipt system. Consequently, they failed to attract a critical mass of business on a regular basis. Further evidence from Eastern and Southern Africa suggests that commodity exchanges in the region had limited success in attracting financial institutions both as an agent for settling payments and, crucially, as a lender to exchange participants (Jayne et al. 2014).

Mbeng Mezui et al. (2013) provide a checklist of good practices critical to the success of a commodity exchange. It proposes a measured role for the government, which must provide the regulatory framework, including for a warehouse receipt system, and funding for the exchange as a shareholder, and demonstration of its commitment to making it work. There is also scope for commodity exchanges to utilise digital technologies and enable transactions in commodity futures.

Agricultural marketing boards

Agricultural marketing boards (AMBs) are state-controlled or state-sanctioned entities vested with quasi-monopoly power over the purchase or sale of agricultural commodities. Once preponderant across Africa, AMBs have waned since the structural adjustment era and rarely active in food markets. (Barrett and Mutambatsere 2008). This may be because agricultural marketing boards often offered a poor deal for farmers, forcing them to accept lower prices than if they sold their produce on the open market (Acemoglu, Johnson and Robinson 2005; Williams 1985). Some governments have said that the 'rents' extracted from farmers are used to fund national development, but this has often not happened (Manley, Heller and Davis 2022, p.38).

It is telling that best examples of the current functioning of AMBs come from the commodity sector rather than the food sector. Studies of cocoa marketing boards in Ghana and Nigeria suggest that these institutions have had a positive impact on cocoa production. In Ghana, the state-run marketing board, COCOBOD, controls all aspects of domestic cocoa marketing and has a de facto monopoly on cocoa exports. However, COCOBOD has demonstrated stewardship, leveraged its strengths in quality control and export management, and implemented effective policies, including a price stabilisation mechanism, that protected farmers' revenues (Matthew et al. 2004). Similar best-practice lessons can be drawn from an analysis of the success of the Nigeria Cocoa Marketing Board. The board focused on productivity improvement and sustainability of the Nigerian cocoa industry, intervening in some unconventional areas, such as disease control, quality assurance and research (Ayinde 2014). The experiences of these AMBs and lessons from the past offer useful insights that other marketing boards can follow.

Multinational market intermediaries

A wide range of multinational businesses exert various degrees of influence on agricultural development and food security on the continent. Their involvement spans the entire agricultural value chain from provision of inputs, machinery, equipment, technology transfer and innovation to investment in agricultural infrastructure, agricultural export processing and marketing. The activities of multinational corporations (MNCs) present both opportunities and challenges, which require careful balancing through strategic partnerships, collaborative solutions, and policy interventions to optimise the benefits that these actors can bring to African agriculture.

In agricultural commodity and food production, MNCs are important suppliers of agricultural inputs such as seeds, fertilisers, pesticides and machinery. Seed and biotechnology companies like Syngenta, DuPont and Bayer lead research and development investments to develop improved seed varieties that are adapted to local conditions. For example, the Water Efficient Maize for Africa project, a partnership between Monsanto and the African Agricultural Technology Foundation, has developed drought-tolerant maize varieties that have shown promising results in countries like Kenya and Uganda (Oikeh et al. 2014). Other MNCs, such as Nutrien, Yara International and BASF, are global suppliers of fertilisers and agrochemicals to African countries and collaborate with African governments and farmer organisations to promote effective fertiliser use across the continent (AFAP 2021). Partnerships between MNCs and local agricultural research institutions also yield context-specific solutions that cater to Africa's unique challenges.

MNCs are further active in agricultural mechanisation, with entities like John Deere and AGCO providing tractors, combine harvesters and other agricultural machinery to African farmers. These technologies can enhance farm productivity and reduce the labour-intensity of agricultural activities (FAO and UNIDO 2008), making them attractive to the youth.

In agricultural processing, marketing and export, MNCs like Olam International have established processing plants across Africa, notably in Ghana and Nigeria (Olam 2021). Local processing activities can reduce post-harvest losses and carry other benefits for producing countries (Urugo et al. 2024). They help to build local capacities to meet sanitary and phytosanitary standards and norms. In the horticultural sector, Syngenta's technologies for pest and disease management have enabled farmers to produce higher-quality and safer products for export, enhancing the reputation and competitiveness of African fruits and vegetables in international markets (Arimond et al. 2013).

However, the involvement of MNCs in African agriculture is not without controversy. One contentious aspect of such involvement has been the acquisition by foreign entities of large tracts of land, or 'land grab', which has raised concerns about land rights, displacement of local communities, and environmental sustainability (Chung and Gagné 2021). While Chinese entities have attracted attention in land deals, entities from other countries have been involved too. For example, land leases by a Saudi Arabian company for the cultivation of rice in a water-scarce region of Ethiopia mainly for export have raised questions about sustainability implications for food security in a country where rice is not widely consumed (Vidal 2010).

MNCs' proprietary control over seeds and biotechnology products can limit farmers' access to critical inputs, perpetuate seed dependency, hinder (sometimes more collaborative) traditional farming practices and undermine agro-biodiversity (Greenberg 2024, p.175; Kloppenburg 2010; Wynberg 2024, p.346). The combination of input market concentration, power imbalances in supply chains, and intellectual property rights provides MNCs with strong advantages over African farmers. African governments have an important role to play in regulating markets, ensuring fair competition and protecting the interests of farmers.

Summary

In CAADP, Africa has a policy blueprint for boosting agricultural development and trade. CAADP requires governments to allocate at least 10 per cent of public expenditure to agriculture and to aim for 6 per cent annual growth in the sector. These goals are reiterated periodically, notably in the 2014 Malabo Declaration. Reviews, however, suggest that only one country – Rwanda – is on track to achieving the CAADP goals. Financial resources remain a major constraint. While there are some good examples of the impact of agricultural financing, there is scope for scaling up private investment, farmers' access to credit, FDI, foreign aid and climate finance. Development partners provide relatively little foreign aid to agricultural development in Africa despite the clear understanding that this sector is critical for achieving international goals on poverty and hunger. Food aid needs to be carefully managed in order not to disincentivise local production. It has been noted that capacities vary among actors and institutions that mediate production, markets and trade such as farmers, 'middlemen', cooperatives, commodity exchanges and agricultural marketing boards.

With the bulk of African agriculture still in the hands of small-scale farmers, any measures to boost investment must necessarily focus on smallholders. However, the rise of contract farming and a class of medium-scale farmers are promising developments especially since this class of farmers have stronger commercial ambitions than the smallholders. Agricultural commercialisation is arguably the most viable pathway for smallholders to increase their output, income and food security, but there are huge challenges as regards imperfect or missing markets and institutions. Alternatively, some smallholder farmers can seek employment outside agriculture (Fan and Rue 2020). Partnerships with MNCs can be beneficial where local interests are well safeguarded.

Notes

[1] Being 'on track' does not mean that the target has already been achieved. Rather, it means that the African Union has assessed that the target would be met by its specified timeline if progress from the baseline to the target is linear (African Union n.d., p.16).

[2] The data is from FAOSTAT. The data is not available consistently for a common period. The averages are computed for the most recent four years for which data is available for a given country.

[3] See, for example, the WTO-OECD Aid for Trade at a Glance 2022 report (OECD and WTO 2022).

[4] The bank has presented this as Aid for Trade (World Trade Organization 2023, pp.1–2).

[5] See *World Agroforestry* (2024).

[6] See IFAD (n.d.).

[7] See ESCAP (n.d.).

[8] Warehouse receipt systems are '[a] process where owners of commodities deposit their commodities in a certified warehouse and are issued with documents known as Warehouse Receipt as proof of ownership' (Warehouse Receipt System(WRS) n.d.).

References

Abebe, Gumataw K.; Bijman, Jos; and Royer, Annie (2016) 'Are Middlemen Facilitators or Barriers to Improve Smallholders' Welfare in Rural Economies? Empirical Evidence from Ethiopia', *Journal of Rural Studies*, vol. 43, pp.203–213. https://doi.org/10.1016/j.jrurstud.2015.12.004

Acemoglu, Daron; Johnson, Simon; and Robinson, James A. (2005) 'Chapter 6 Institutions as a Fundamental Cause of Long-Run Growth', in Aghion, P. and Durlauf, S. N. (eds) *Handbook of Economic Growth*, Elsevier, pp.385–472. https://doi.org/10.1016/S1574-0684(05)01006-3

ActionAid (2013). 'Fair Shares: Is CAADP Working?' Johannesburg: ActionAid. https://perma.cc/GHH4-P4RL

ActionAid (2015) 'Contract Farming and Out-Grower Schemes Appropriate Development Models to Tackle Poverty and Hunger?', Policy Discussion Paper. https://perma.cc/39X6-RBDH

Adenle, Ademola A.; Wedig, Karin; and Azadi, Hossein (2019) 'Sustainable Agriculture and Food Security in Africa: The Role of Innovative Technologies and International Organizations', *Technology in Society*, vol. 58, p.101143. https://doi.org/10.1016/j.techsoc.2019.05.007

AFAP (African Fertilizer and Agribusiness Partnership) (2021) 'About Us'. https://afap-partnership.org/about/

AfDB (2022) *African Economic Outlook 2022: Supporting Climate Resilience and a Just Energy Transition in Africa*, Abidjan: African Development Bank. https://www.afdb.org/en/documents/african-economic-outlook-2022

African Union (n.d.) *3rd CAADP Biennial Review Report*. Addis Ababa: African Union. https://perma.cc/26P4-J9L5

AgFunder (2022) *2022 Africa AgriFoodTech Investment Report*. https://perma.cc/7WJE-5YEH

AGRA (2022) *Empowering Africa's Food Systems for the Future. Africa Agriculture Status Report 2023*. Nairobi: AGRA.

Ahluwalia, Montek Singh and Patel, Utkarsh (2022) 'Financing Climate Change Mitigation and Adaptation in Developing Countries', in A. Bhattacharya, H. Kharas, and J. W. McArthur (eds) *Keys to Climate Action*. The Brookings Institution, pp.309–331. https://perma.cc/84ZF-DV7P

Alabi, Reuben A. (2014) 'Impact of Agricultural Foreign Aid on Agricultural Growth in Sub-Saharan Africa: A Dynamic Specification', AGRODEP Working Paper 0006, Washington, DC: IFPRI. https://hdl.handle.net/10568/149556

Arimond, M.; Hawkes, C.; Ruel, M. T.; Sifri, Z.; Berti, P. R.; Leroy, J. L., Low, J. W., Brown, L. R. and Frongillo, E. A. et al. (2013) 'Agricultural Interventions and Nutrition: Lessons from the Past and New Evidence', *Advances in Nutrition*, vol. 4, no. 6, pp.749–54. https://doi.org/10.1079/9781845937140.0041

Arouna, Aminou; Michler, Jeffrey D.; and Lokossou, Jourdain C. (2021) 'Contract Farming and Rural Transformation: Evidence from a Field Experiment in Benin', *Journal of Development Economics*, vol. 151, 102626. https://doi.org/10.1016/j.jdeveco.2021.102626

Ayinde, O. (2014) 'Is Marketing Board a Barrier or a Stimulant of Agricultural Production in West Africa? A Comparative Study of Ghana and Nigeria Cocoa Pricing Eras', *Ghana Journal of Development Studies*, vol. 11, no. 2, pp.50–66. https://doi.org/10.4314/gjds.v11i2.4

Badiane, Ousmane; Collins, Julia; and Ulimwengu, John M. (2020) 'The Past, Present, and Future of Agriculture Policy in Africa', in Resnick, D.; Diao, X.; and Tadesse, G. (eds) *2020 Annual Trends and Outlook Report: Sustaining Africa's Agrifood System Transformation: The Role of Public Policies.* Washington, DC and Kigali: IFPRI and AKADEMIYA2063, pp.9–25. https://doi.org/10.2499/9780896293946_02

Banerjee, Abhijit; Duflo, Esther; Glennerster, Rachel; and Kinnan, Cynthia (2015) 'The Miracle of Microfinance? Evidence from a Randomized Evaluation', *American Economic Journal: Applied Economics*, vol. 7, no. 1, pp.22–53. https://doi.org/10.1257/app.20130533

Barrett, Christopher B.; and Maxwell, Dan (2007) *Food Aid After Fifty Years: Recasting its Role.* London: Routledge. https://doi.org/10.4324/9780203799536

Barrett, Christopher B.; and Mutambatsere, Emelly (2008) 'Marketing Boards', in Blume, L. E. and Durlauf, S. N. (eds) *The New Palgrave Dictionary of Economics.* 2nd edition, London: Palgrave Macmillan.

Beintema, Nienke; and Stads, Gert-Jan (2017) 'A Comprehensive Overview of Investments and Human Resource Capacity in African Agricultural Research', ASTI Synthesis Report, Washington, DC: IFPRI. https://perma.cc/QB7Q-DZ62

Béné, Christophe; Devereux, Stephen; and Sabates-Wheeler, Rachel (2012) 'Shocks and Social Protection in the Horn of Africa: Analysis from the Productive Safety Net Programme in Ethiopia', IDS Working Paper, 2012(395), pp.1–120. https://perma.cc/G74F-34NV

Benin, Samuel (2018) 'From Maputo to Malabo: How Has CAADP Fared?', ReSAKSS Working Paper 40, Washington, DC: IFPRI. https://perma.cc/659M-54JZ

Bernard, Tanguy; and Taffesse, Alemayehu S. (2012) 'Returns to Scope? Smallholders' Commercialisation through Multipurpose Cooperatives in Ethiopia', *Journal of African Economies*, vol. 21, no. 3, pp.440–64. https://doi.org/10.1093/jae/ejs002

Biscaye, Pierre E.; Reynolds, Travis W.; and Anderson, C. Leigh (2017) 'Relative Effectiveness of Bilateral and Multilateral Aid on Development Outcomes', *Review of Development Economics*, vol. 21, no. 4, pp.1425–47. https://doi.org/10.1111/rode.12303

Brüntrup, Michael (2011) 'African Developments: The Comprehensive Africa Agriculture Development Programme (CAADP) Is an Opportunity for African Agriculture', Briefing Paper 4/2011, Bonn: German Development Institute. https://perma.cc/D8G3-MGHV

Castell, H. (2019) 'New Investment Funds Impact African Agriculture', *Blended Finance*, vol. 193 (June–August), pp.4–7.

Chung, Youjin.; and Gagné, Marie (2021) 'What Happened to Land Grabs in Africa?', *Africa Is a Country*, 13 October. https://perma.cc/6KGX-LAX9

Cordaid (2021) 'Breaking Down the Barriers for African Smallholder Farmers', Cordaid. https://perma.cc/M6KQ-DFX8

Cuevas, Carlos; and Pagura, Maria (2016) *Agricultural Value Chain Finance A Guide for Bankers*. World Bank Group. https://www.mfw4a.org/sites/default/files/resources/Bankers_Guide _to_AVCF.pdf

De Brauw, Alan; and Bulte, Erwin (2021) 'The Evolution of Agricultural Value Chains in Africa', in De Brauw, A. and Bulte, E. (eds) *African Farmers, Value Chains and Agricultural Development: An Economic and Institutional Perspective*, Cham: Springer.

Diallo, Mariam; and Wouterse, Fleur (2023) 'Agricultural Development Promises More Growth and Less Poverty in Africa: Modelling the Potential Impact of Implementing the Comprehensive Africa Agriculture Development Programme in Six Countries', *Development Policy Review*, vol. 41, no. 3. https://doi.org/10.1111/dpr.12669

Dogan, Berna (2022) 'Does FDI in Agriculture Promote Food Security in Developing Countries? The Role of Land Governance', *Transnational Corporations*, vol. 22, no. 2, pp.47–74. https://perma.cc/6FCY-6SNT

Economic Commission for Africa (2017) *Expanding and Strengthening Local Entrepreneurship for Structural Transformation in Africa*, Addis Ababa: Economic Commission for Africa. https://perma.cc/N9JN-NZTW

Economic Commission for Africa (2019) *Financial Regulation for Inclusive Growth in Africa*, Addis Ababa: United Nations Economic Commission for Africa. https://perma.cc/LGM4-D7DP

Economic Commission for Africa; African Development Bank Group; and African Union Commission (2021) *African Statistical Yearbook 2021 = Annuaire Statistique pour l'Afrique 2021*, Addis Ababa: Economic Commission for Africa. https://www.afdb.org/en/documents/african-statistical-yearbook-2021

The Economist (2022) 'Middlemen Are the Invisible Links in African Agriculture', 1 January. https://perma.cc/F9NP-FCZM

Eleta, V. (2020). 'In Defense of Agricultural Middlemen', LinkedIn, 4 December 2020. https://perma.cc/VMJ7-7XTJ

Enete, A. A. (2009) 'Middlemen and Smallholder Farmers in Cassava Marketing in Africa', *Tropicultura*, vol. 27, no. 1, pp.40–44. http://www.tropicultura.org/text/v27n1/40.pdf

ESCAP (n.d.) 'Catalyzing Women's Entrepreneurship', ESCAP.
    https://www.unescap.org/projects/cwe

Fan, Shenggen; and Rue, Christopher (2020) 'The Role of Smallholder
    Farms in a Changing World', in Gomez y Paloma, Sergio, Riesgo,
    Laura and Louhichi, Kamel (eds) *The Role of Smallholder Farms
    in Food and Nutrition Security*, Cham: Springer, pp.13–28.
    https://doi.org/10.1007/978-3-030-42148-9_2

FAO (2010) *Promoting Employment and Entrepreneurship for
    Vulnerable Youths in West Bank and the Gaza Strip*, Rome: FAO.
    https://perma.cc/YCN6-CUCW

FAO (2014) 'Grave Food Security Concerns following the Ebola Outbreak in
    Liberia, Sierra Leone and Guinea', Special Alert No. 333, Global Informa-
    tion and Early Warning System on Food and Agriculture, 2 September.

FAO (2018a) *The State of Agricultural Commodity Markets 2018.
    Agriculture, climate change and food security*, Rome: FAO.
    http://www.fao.org/3/I9542EN/i9542en.pdf

FAO (2018b) *Agricultural Investment Funds for development: Descriptive
    Analysis and Lessons Learned from Fund Management, Performance, and
    Private-Public Collaboration*, Rome: FAO.

FAO (2021) *Trade and Food Safety Standards: African Free Trade and Food
    Safety*, Rome: FAO.

FAO; and UNIDO (2008) *Agricultural Mechanization in Africa… Time for
    Action Planning Investment for Enhanced Agricultural Productivity Report
    of an Expert Group Meeting January 2008, Vienna, Austria*, Rome and
    Vienna: FAO and UNIDO. https://www.fao.org/4/k2584e/k2584e.pdf

Feyertag, Joseph; and Bowie, Ben (2021) 'The Financial Costs of Mitigating
    Social Risks: Costs and Effectiveness of Risk Mitigation Strategies
    for Emerging Market Investors', ODI Report, London: ODI.
    https://odi.org/en/publications/the-financial-costs-of-mitigating-social
    -risks-costs-and-effectiveness-of-risk-mitigation-strategies-for-emerging
    -market-investors

Food and Agriculture Organization of the United Nations (n.d.) 'FAOSTAT'.
    https://perma.cc/5Y7P-WM75

Gerlach, Ann-Christin; and Liu, Pascal (2010) 'Resource-Seeking Foreign
    Direct Investment in African Agriculture: A Review of Case Studies',
    FAO Commodity and Trade Policy Research Working Paper, no. 31,
    September. https://perma.cc/CGM9-ZXE5

Goedde, Lutz; Ooko-Ombaka, Amandla; and Pais, Gillian (2019) 'Winning
    in Africa's Agricultural Market', McKinsey & Company, 15 February.
    https://perma.cc/GM8R-WED3

Greenberg, Stephen (2024) 'Corporate Expansion in African Seed Systems: Implications for Agricultural Biodiversity and Food Sovereignty', in Wynberg, R. (ed.) *African Perspectives on Agroecology*, p.165.

Grow Further (2022) 'Investors Are Finally Paying Attention to African Agriculture', Grow Further | Connecting People and Ideas for a Food-Secure Future. https://perma.cc/J5YP-LBQP

Gulrajani, Nilima (2016) 'Bilateral versus Multilateral Aid Channels Strategic choices for donors', ODI. https://perma.cc/2SEG-9GR5

Hall, Ruth, Scoones, Ian and Dzodzi Tsikata (2017) 'Plantations, outgrowers and commercial farming in Africa: agricultural commercialisation and implications for agrarian change', *The Journal of Peasant Studies*, 44(3), pp.515–537. https://doi.org/10.1080/03066150.2016.1263187

Harou, Aurélie. P.; Upton, Joanna B.; Lentz, Erin C.; Barrett, Christopher B.; and Gómez Miguel I. (2013) 'Tradeoffs or Synergies? Assessing Local and Regional Food Aid Procurement through Case Studies in Burkina Faso and Guatemala', *Impacts of Innovative Food Assistance Instruments*, vol. 49, pp.44–57. https://doi.org/10.1016/j.worlddev.2013.01.020

Harvey, Fiona (2022) 'World Bank Criticised over Climate Crisis Spending', *The Guardian*, 3 October. https://perma.cc/2RS8-FTGS

Husmann, Christine and Kubik, Zaneta (2019) 'Foreign Direct Investment in the African Food and Agriculture Sector: Trends, Determinants and Impacts', *SSRN Electronic Journal* [Preprint]. https://doi.org/10.2139/ssrn.3370799

IAASTD (International Assessment of Agricultural Knowledge, Science and Technology for Development) (2009) *Agriculture at a Crossroads: Global Report*. Washington, DC: Island Press.

IFAD (n.d.) 'Future of Food Security in Africa: How Innovation Can Help Our Future in Times of a Global Food Crisis?', IFAD. https://www.ifad.org/en/w/events/future-of-food-security-in-africa-how-innovation-can-help-our-future-in-times-of-a-global-food-crisis-

Jayne, T. S.; Chamberlin, Jordan; Traub, Lulama; Sitko, Nicholas; Muyanga, Milu; Yeboah, Felix K.; Anseeuw, Ward; Chapoto, Antony et al. (2016) 'Africa's Changing Farm Size Distribution Patterns: The Rise of Medium-Scale Farms', *Agricultural Economics*, vol. 47, no. S1, pp.197–214. https://doi.org/10.1111/agec.12308

Jayne, T. S.; Sturgess, Chris.; Kopicki, Ron; and Sitko, Nicholas (2014) 'Agricultural Commodity Exchanges and the Development of Grain Markets and Trade in Africa: A Review of Recent Experience', Working Paper no. 88, Indaba Agricultural Policy Research Institute, October.

Jere, Paul (2007) 'The Impact of Food Aid on Food Markets and Food Security in Malawi', EQUINET Discussion Paper no. 45, April.

Kapuya, Tinashe; Mutyasira, Vine; Haddad, Lawrence; and Keizire, Boaz B (2022) 'A Stocktake of Africa's Food Systems', in AGRA (ed.) *Empowering Africa's Food Systems for the Future. Africa Agriculture Status Report 2023*, Nairobi: AGRA.

Kloppenburg, J. (2010) 'Impeding Dispossession, Enabling Repossession: Biological Open Source and the Recovery of Seed Sovereignty', *Journal of Agrarian Change*, vol. 10, no. 3, pp.367–88. https://doi.org/10.1111/j.1471-0366.2010.00275.x

Kumar, Sanjay (2015) 'Green Climate Fund Faces Slew of Criticism. First Tranche of Aid Projects Prompts Concern over Operations of Fund for Developing Nations', *Nature*, vol. 527, pp.419–20. https://doi.org/10.1038/nature.2015.18815

Larson, Donald F.; Otsuka, Keijiro; Matsumoto, Tomoya; and Kilic, Talip (2014) 'Should African Rural Development Strategies Depend on Smallholder Farms? An Exploration of the Inverse-Productivity Hypothesis', *Agricultural Economics*, vol. 45, pp.355–67. https://doi.org/10.1111/agec.12070

Lay, Jann; Anseeuw, Ward; Eckert, Sandraorcid-logo; Flachsbarth, Insa; Kubitza, Christoph; Nolte, Kerstin; Giger, Markus (2021) *Taking Stock of the Global Land Rush: Few Development Benefits, Many Human and Environmental Risks. Analytical Report III*, Centre for Development and Environment, University of Bern; Centre de coopération internationale en recherche agronomique pour le développement; German Institute of Global and Area Studies; University of Pretoria; Bern Open Publishing. https://doi.org/10.48350/156861

Lentz, Erin C.; Passarelli, Simone; and Barrett, Christopher B. (2013) 'The Timeliness and Cost-Effectiveness of the Local and Regional Procurement of Food Aid', *Impacts of Innovative Food Assistance Instruments*, vol. 49, pp.9–18. https://doi.org/10.1016/j.worlddev.2013.01.017

Maina, Nyaguthii (2022) 'Realizing Responsible Investment in Agriculture: What Can Policymakers in Africa Learn from Southeast Asia?', *Policy Analysis*, 27 July. IISD. https://perma.cc/EM9N-QSXH

Makombe, Tsitsi; and Kurtz, Julie (2020) 'Second Biennial Review Report Highlights Urgent Need to Accelerate Progress toward Achieving the Malabo Declaration Goals by 2025', *IFPRI blog*, 17 February. https://perma.cc/A5F8-X6ME

Malhotra, Swati; and Vos, Rob (2021) 'Africa's Processed Food Revolution and the Double Burden of Malnutrition', *IFPRI blog*, 11 March. https://perma.cc/N3C6-JNMH

Manley, David; Heller, Patrick R. P.; and Davis, William (2022) *No Time to Waste: Governing Cobalt Amid the Energy Transition*, Natural Resource Governance Institute. https://perma.cc/V3B7-9PZL

Markandya, Anil; and González-Eguino, Mikel (2019) 'Integrated Assessment for Identifying Climate Finance Needs for Loss and Damage: A Critical Review', in Mechler, R. et al. (eds) *Loss and Damage from Climate Change: Concepts, Methods and Policy Options*, Cham: Springer International Publishing, pp.343–62. https://doi.org/10.1007/978-3-319-72026-5_14

Matthew, Uwakonye; Nazemzadeh, Asghar; Gbolahan Solomon, Osho; and Etundi, William J. (2004) 'Social Welfare Effect of Ghana Cocoa Price Stabilization: Time-Series Projection and Analysis', *International Business & Economics Research Journal*, vol. 3, no. 12, pp.45–54. https://doi.org/10.19030/iber.v3i12.3741

Mbeng Mezui, Cedric Achille; Rutten, Lamon; Sekioua, Sofiane; Zhang, Jian; N'Diaye, Max Magor; Kabanyane, Nontle; Arvanitis, Yannis; and Duru, Uche et al. (2013) *Guidebook on African Commodity and Derivatives Exchanges*, Tunis: African Development Bank. https://www.afdb.org/fileadmin/uploads/afdb/Documents/Publications/Guidebook_on_African_Commodity_and_Derivatives_Exchanges.pdf

McArthur, John; and Sachs, Jeffrey D. (2018) 'Agriculture, Aid and Economic Growth in Africa', World Bank Policy Research Working Paper No. 8447. https://ssrn.com/abstract=3182877

Meemken, Eva-Marie; and Bellemare, Marc F. (2019) 'Smallholder Farmers and Contract Farming in Developing Countries', *Proceedings of the National Academy of Sciences*, vol. 117, no. 1, pp.259–64. https://doi.org/10.1073/pnas.1909501116

Mengoub, Fatima Ezzahra (2018) 'Agricultural Investment in Africa: A Low Level... Numerous Opportunities', Policy Brief PB-18/02, January, OCP Policy Center. https://perma.cc/W9HP-KZM5

Mercier, Stephanie (2020) 'The Emerging Role of Cooperatives in African Agriculture', *AGWEB Farm Journal*, 2 July. https://www.agweb.com/opinion/emerging-role-cooperatives-african-agriculture

Minot, Nicholas W. (2015) 'Contract Farming in Sub-Saharan Africa: Opportunities and Challenges', Mimeo, IFPRI.

Mitchell, Chris (2019) 'Africa's Food System Needs a Better Class of Middlemen', *World Economic Forum*, 1 September.

https://www.weforum.org/stories/2019/09/africas-food-system-needs
-a-better-class-of-middlemen/

Mogues, Tewodaj; Fan, Shenggen; and Benin, Samuel (2015) 'Public Investments in and for Agriculture', *European Journal of Development Research*,
vol. 27, no. 3, pp.337–52. https://doi.org/10.1057/ejdr.2015.40

Muyanga, Milu; and Jayne, Thomas (2018) 'Medium-Scale Farms Are
on the Rise in Africa. Why This Is Good News', *The Conversation*.
https://perma.cc/6XFC-SAHB

OECD (2006) *The Development Effectiveness of Food Aid: Does Tying Matter?*,
Paris: OECD Publishing. https://doi.org/10.1787/9789264013476-en

OECD; and FAO (2022) *OECD-FAO Agricultural Outlook 2022–2031*,
Paris: OECD Publishing. https://doi.org/10.1787/9789264013476-en

OECD; and WTO (2022) *Aid for Trade at a Glance 2022: Empowering
Connected, Sustainable Trade*, Paris: OECD Publishing.
https://doi.org/10.1787/9ce2b7ba-en

Oikeh, Sylvester; Ngonyamo-Majee, Dianah; Mugo, Stephen I. N.;
Mashingaidze, Kingstone; Cook, Vanessa; and Stephens, Michael (2014)
'The Water Efficient Maize for Africa Project as an Example of a
Public–Private Partnership', in Songstad, D. D., Hatfield, J. L. and
Tomes, D. T. (eds) *Convergence of Food Security, Energy Security
and Sustainable Agriculture*, Berlin, Heidelberg: Springer, pp.317–29.
https://doi.org/10.1007/978-3-642-55262-5_13

Organisation for Economic Co-operation and Development (n.d.) 'Food
Aid', *OECD*. https://www.oecd.org/en/data/indicators/food-aid.html

Olam (2021) 'About Olam'. https://www.olamgroup.com/about-us.html

Oyewole, Babafemi (2022) 'Boosting Smallholder Farmers' Productivity
to Feed Africa against the Looming Food Crisis', Keynote
address at the AfDB Virtual Evaluation Week, 28–29 September.
https://idev.afdb.org/sites/default/files/documents/files/Evaluation%
20Week%202022%20-%20BOOSTING%20SMALLHOLDER%20
FARMERS_Dr%20Babafemi_Agrulture%20session%20%281%29.pdf

Parlasca, Martin C.; Johnen, Constantin; and Qaim, Matin (2022)
'Use of Mobile Financial Services among Farmers in Africa:
Insights from Kenya', *Global Food Security*, vol. 32, p.100590.
https://doi.org/10.1016/j.gfs.2021.100590

Phatisa (2021) 'Homepage'. https://perma.cc/BFL6-DMTC

Poku, Adu-Gyamfi; Birner, Regina; and Gupta, Saurabh (2018) 'Making
Contract Farming Arrangements Work in Africa's Bioeconomy: Evidence

from Cassava Outgrower Schemes in Ghana', *Sustainability*, vol. 10, no. 5, p.1604. https://doi.org/10.3390/su10051604

Rashid, Shahidur; Winter-Nelson, Alex; and Garcia, Philip (2010) 'Purpose and Potential for Commodity Exchanges in African Economies', IFPRI Discussion Paper 01035, Washington, DC: International Food Policy Research Institute. https://perma.cc/2LHB-W4RQ

Robbins, Peter; and Catholic Relief Services (2011) *Commodity Exchanges and Smallholders in Africa*, London: International Institute for Environment and Development/Sustainable Food Lab. https://perma.cc/LY88-VJX5

Savoy, Conor M. (2022) 'Access to Finance for Smallholder Farmers', Center for Strategic and International Studies, 7 December. https://perma.cc/2LTD-HRUG

'SDG Indicators' (n.d.). fao.org: FAOSTAT. https://perma.cc/EAQ3-R84N

Shikuku, Kelvin M.; Valdivia, Roberto O.; Paul, Birthe K.; Mwongera, Caroline; Winowiecki, Leigh; Läderach, Peter; Herrero, Mario; and Silvestri, Silvia (2017) 'Prioritizing Climate-Smart Livestock Technologies in Rural Tanzania: A Minimum Data Approach', *Agricultural Systems*, vol. 151, pp.204–16. https://doi.org/10.1016/j.agsy.2016.06.004

Sifa, Chiyoge B. (2014) 'Role of Cooperatives in Agricultural Development and Food Security in Africa'. https://perma.cc/GNA8-VNZL

Signé, Landry (2017) 'The Quest for Food Security and Agricultural Transformation in Africa: Is the CAADP the Answer?', *Africa in Focus*, 6 October. https://perma.cc/KMT6-P9U6

Sikwela, M. M.; Fuyane, N.; and Mushunje, A. (2016) 'The Role of Cooperatives in Empowering Smallholder Farmers to Access Markets: A Case Study of Eastern Cape and KwaZulu Natal Cooperatives in South Africa', *International Journal of Development and Sustainability*, vol. 5, no. 11, pp.536–52. https://perma.cc/A75W-ZMNK

SME Finance Forum (2017) *Value Chain Financing*. https://www.smefinanceforum.org/tool/tools/value-chain-financing

Songwe, Vera (2011) 'From Subsistence Agriculture to Agribusiness in Africa: The Role of Commodity Exchanges', Brookings, 5 October. https://perma.cc/9TE9-4JED

Ssozi, John; Asongu, Simplice; and Amavilah, Voxi H. (2019) 'The Effectiveness of Development Aid for Agriculture in Sub-Saharan Africa', *Journal of Economic Studies*, vol. 46, no. 2, pp.284–305. https://doi.org/10.1108/JES-11-2017-0324

Stewart, Alana (2020) 'The Impact of Microfinance on Smallholder Farming Households in Africa: Evidence from Zambia', IOA. https://perma.cc/D6HS-4R9E

Transparency International (2022) *Corruption-Free Climate Finance: Strengthening multilateral funds.* Transparency International. https://perma.cc/9GZ2-SWGB

Ulimwengu, John M.; Nwafor, Apollos; and Nhlengethwa, Sibusiso (2022) 'Assessing Structural Failure of African Food Systems', in AGRA (ed.) *Empowering Africa's Food Systems for the Future. Africa Agriculture Status Report 2023*, Nairobi: AGRA, pp.27–51. https://perma.cc/D5DR-EVNQ

UNFCCC (2023) 'COP28 Agreement Signals "Beginning of the End" of the Fossil Fuel Era United Nations Climate Change'. https://perma.cc/JXH4-25NQ

UN Women (2020) 'Women Cooperatives Boost Agriculture and Savings in Rural Ethiopia', UN Women Africa. https://perma.cc/VD7Y-KPJX

United Nations Conference on Trade and Development (2023) *World Investment Report 2023 Investing in Sustainable Energy for All*, Geneva: United Nations. https://perma.cc/K8DN-FRY5

Urugo, Markos Makiso; Yohannis, Eyasu; Teka, Tilahun A.; Gemede, Habtamu Fekadu; Tola, Yetenayet B.; Forsido, Sirawdink Fikreyesus; Tessema, Ararsa; and Suraj, Mohammed et al. (2024) 'Addressing Post-Harvest Losses through Agro-processing for Sustainable Development in Ethiopia', *Journal of Agriculture and Food Research*, vol. 18, p.101316. https://doi.org/10.1016/j.jafr.2024.101316

van Rooyen, C.; Stewart, R.; and de Wet, T. (2012) 'The Impact of Microfinance in Sub-Saharan Africa: A Systematic Review of the Evidence', *World Development*, vol. 40, no. 11, pp.2249–62. https://doi.org/10.1016/j.worlddev.2012.03.012

Vidal, John (2010) 'How Food and Water Are Driving a 21st-Century African Land Grab', *The Observer*, 7 March. https://perma.cc/735S-8BWN

'Warehouse Receipt System (WRS)' (n.d.) NCPB. https://perma.cc/85HF-EJL8

Watson, Charlene; and Schalatek, Liane (2020) *The Global Climate Finance Architecture*, Climate Finance Fundamentals. https://perma.cc/W6YE-AF6C

Williams, Gavin (1985) 'Marketing without and with Marketing Boards: The Origins of State Marketing Boards in Nigeria', *Review of African Political Economy*, vol. 34, pp.4–15. https://www.scienceopen.com/hosted-document?doi=10.1080/03056248508703647

WEF (World Economic Forum) (2016) *Grow Africa: Partnering to Achieve African Agriculture Transformation*, Geneva: World Economic Forum. https://perma.cc/U97K-XSVE

World Agroforestry (2024) 'Reversing Land Degradation in Africa by Scaling-Up Evergreen Agriculture (Regreening Africa)', World Agroforestry. https://perma.cc/44NP-2VH6

World Bank (2007) *World Development Report 2008: Agriculture for Development*, Washington, DC: The World Bank. http://hdl.handle.net/10986/5990

World Bank (2018) 'Improving Access to Finance for SMEs: Opportunities through Credit Reporting, Secured Lending and Insolvency Practices', Working Paper 129283, Washington, DC: World Bank Group. https://perma.cc/8GSQ-7GNC

World Food Programme (2021a) 'WFP – Saving Lives, Preventing Famine', 15 November. https://www.wfp.org/stories/wfp-saving-lives-preventing-famine

World Food Programme (2021b) *2021 School Feeding Programme Factsheet – WFP Malawi*, Rome: World Food Programme. https://docs.wfp.org/api/documents/WFP-0000131157/download/?_ga =2.28728700.1528062675.1701454596-575233052.1701454595

World Trade Organization (2023) 'Committee on Trade and Development Fifty-Eighth Session on Aid for Trade Note on the Meeting of 12 May 2023', World Trade Organization. https://docs.wto.org/dol2fe/Pages/SS/directdoc.aspx?filename=q:/WT/ COMTD/AFTM58.pdf&Open=True

Wynberg, Rachel (2024) 'Conclusion: Towards Seed and Knowledge Justice for Agroecology', in Wynberg, R. (ed.) *African Perspectives on Agroecology Why Farmer-Led Seed and Knowledge Systems Matter*, Rugby: Practical Action Publishing. http://doi.org/10.3362/9781780447445

5. Intra-African food trade

David Luke, William Davis and Vinaye Dey Ancharaz

This chapter, on the intra-African food trade, builds upon the synopsis presented in Chapter 2 that situated Africa's agriculture and food trade in Africa's overall trade. Intra-African exports are second only to those to the European Union (EU) in importance as a market for exports of food and agricultural commodities and are dominated by trade in food products, while involving less trade in agricultural commodities. In contrast, agricultural commodities comprise a large share of Africa's agricultural exports to countries in Asia, the Americas and elsewhere in the world. Intra-African trade also includes a large informal component in which trade in food products is correspondingly dominant.

This chapter further builds on Chapter 3 on production and consumption of the basic foods. Eight products that make up the 'basic foods' basket were identified: cassava, yams, rice, maize, wheat, meat, poultry and fish. While trade was not the focus of that chapter, the general underperformance of production of most of these foods in relation to global output provided insights into the underlying dynamics of Africa's status as a net food importer. A major implication is that intra-African trade, although dominated by trade in food products, remains relatively small. This is why, as we saw in Chapter 4, agricultural transformation is the overriding objective of the Comprehensive African Agriculture Development Programme (CAADP) and boosting the food component of intra-African trade is a specific Malabo Declaration commitment. The African Continental Free Trade Area (AfCFTA), which came into force in 2019, is an even more ambitious effort to increase trade flows, including on food, within the continent. Intra-African food trade and its composition, regional patterns and informal trade are the focus of this chapter. The likely impact of the AfCFTA is examined in Chapter 6.

Comprising three main sections, the chapter reviews the overall trends in intra-African food trade, followed by a focus on food trade at the regional level including the trade patterns of the basic foods and concluding with an outline of the main features of informal cross-border food trade.

How to cite this book chapter:

Luke, David; Davis, William and Ancharaz, Vinaye Dey (2025) 'Intra-African food trade', in: Luke, David (ed) *How Africa Eats: Trade, Food Security and Climate Risks*, London: LSE Press, pp. 107–124. https://doi.org/10.31389/lsepress.hae.e
License: CC-BY-NC 4.0

5.1 Trends in intra-African food trade

The value of intra-African imports of basic foods (and conversely exports) have grown over the last 10 years, albeit with fluctuations as shown in Figure 5.1. Growth in intra-African imports of basic food tails off after 2015. This appears to be only partly explained by trends in the average price per kg of basic food (either within individual foods or due to a shift to more expensive foods – see Figure 5.2). A rapid rise in demand for fish up to 2015 (as shown in Figure 5.6), which also tails off after 2015, could also explain the trend. This could be linked to a boom in prices of commodities that Africa exports that lasted from 2004 until 2014, which may have supported higher consumption of fish on the continent linked to higher incomes for some persons, with prices bottoming out in 2016 (Cust and Zeufack 2023, p.101; International Monetary Fund 1992).

Cereals, vegetables and fruits, and fish and fish preparations are the major intra-Africa food imports (see Figure 5.3).[1] However, African countries import almost twice as many cereals from the rest of the world than they do from each other. Figure 5.4 provides a breakdown of intra-Africa cereal imports compared to cereal imports from the rest of the world.

Figure 5.1: Intra-African imports of basic foods (US$ billion at 2022 prices), 2012–2021, and average price per kg of traded food (US$)

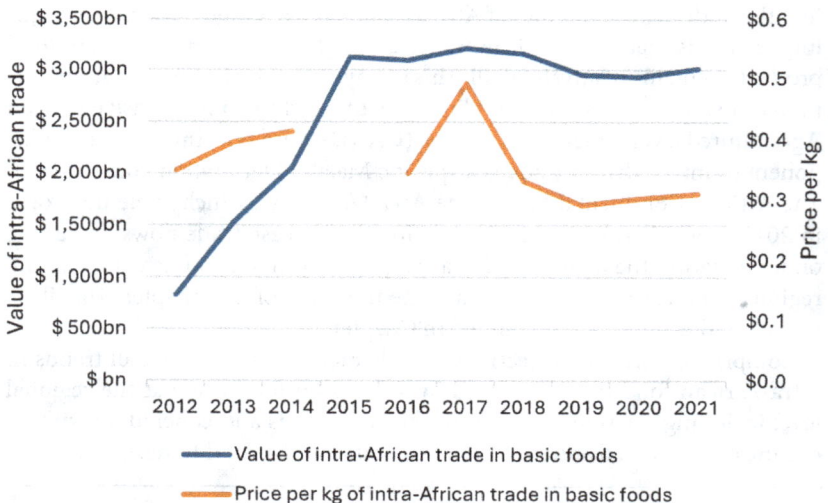

Source: Authors' calculations based on UN Comtrade and GDP deflator (base year varies by country) (2023). Owing to fewer countries reporting in 2022 than in 2021, we present only data up to 2021. Data on traded volumes is not available for all countries in 2015, which is why data on average prices is not available for that year.

Figure 5.2: Share (in percentage) of basic foodstuffs in total intra-African trade in basic foods, by value, 2012–2021

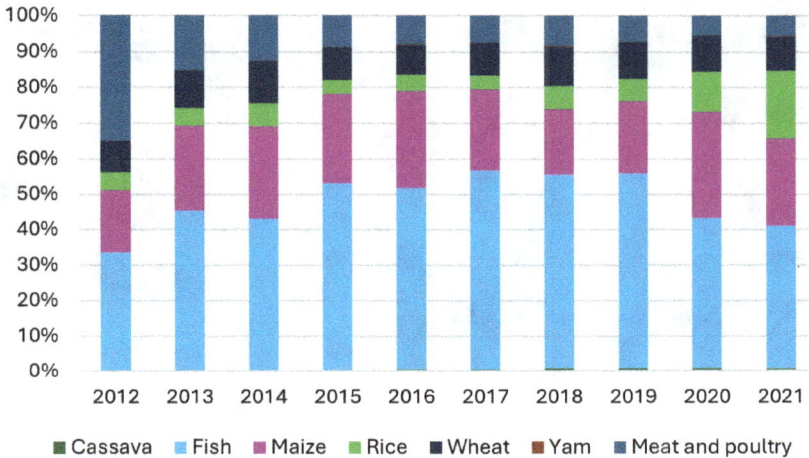

Source: Authors' calculations based on based on UN Comtrade.
Note: For each basic foodstuff shown in this chart, shares in intra-African basic food trade include the contribution of products derived from that basic food. The only exception is yams.

Figure 5.3: Intra-African and world imports of food products by group (in percentage), 2017–2021 averages, by value

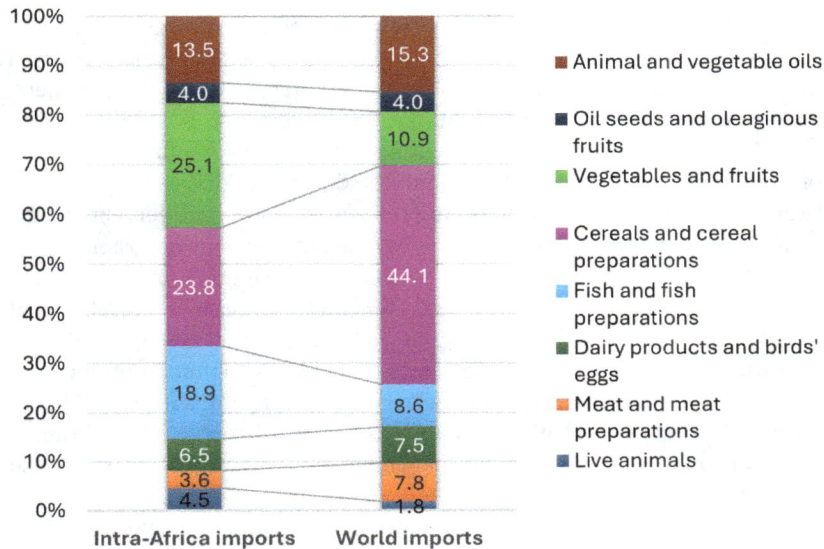

Source: Authors' calculations using data from UNCTADSTAT.

Figure 5.4: Shares (in percentage) of various cereals in Africa's overall imports of cereals, intra-Africa vs. total imports, 2017–2021 averages

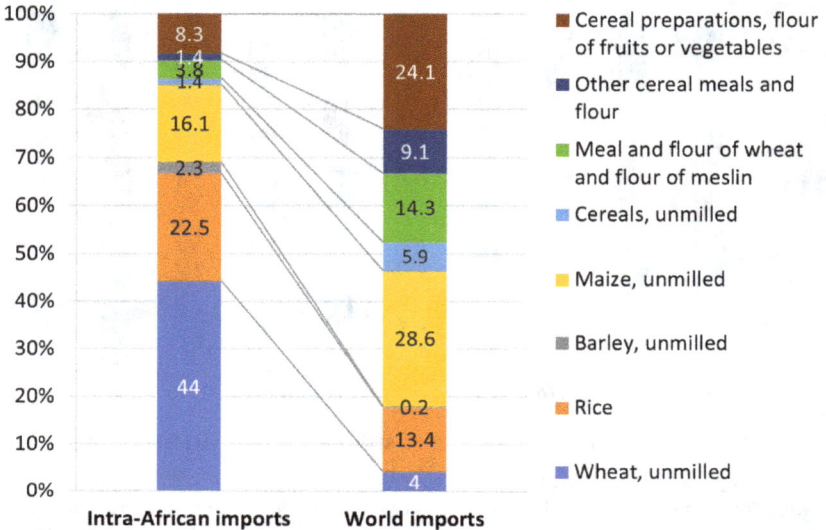

Source: Authors' calculations using data from UNCTADSTAT.

5.2 Regional food trade

Intra-African trade mainly occurs within the regional economic communities rather than between them. The AU recognises eight of these regional organisations as building blocks of economic integration on the continent. Four of the eight (the Common Market for Eastern and Southern African (COMESA), the East African Community (EAC), the Economic Community of West African States (ECOWAS) and the Southern African Development Community (SADC)) have working preferential trade arrangements in the form of free trade areas and have active trade development programmes. Of these, COMESA, EAC and ECOWAS also have customs unions. The Arab Maghreb Union (AMU), the Intergovernmental Authority on Development (IGAD) and the Community of Sahel–Saharan States (CEN-SAD) do not have their own regional trade agreements, while the Economic Community of Central African States (ECCAS) does but it is not fully implemented (United Nations Economic Commission for Africa et al. 2019, p.11; Rettig, Kamau and Muluvi 2023). Figure 5.5 shows membership of the continent's regional economic communities.

As noted in Chapter 3, CAADP requires the regional economic communities (RECs) to have regional agriculture investment programmes. To this end, the RECs are also included in the mutual accountability framework for reporting on progress required by the Malabo Declaration. This recognised

Figure 5.5: Membership of Africa's regional economic communities by country

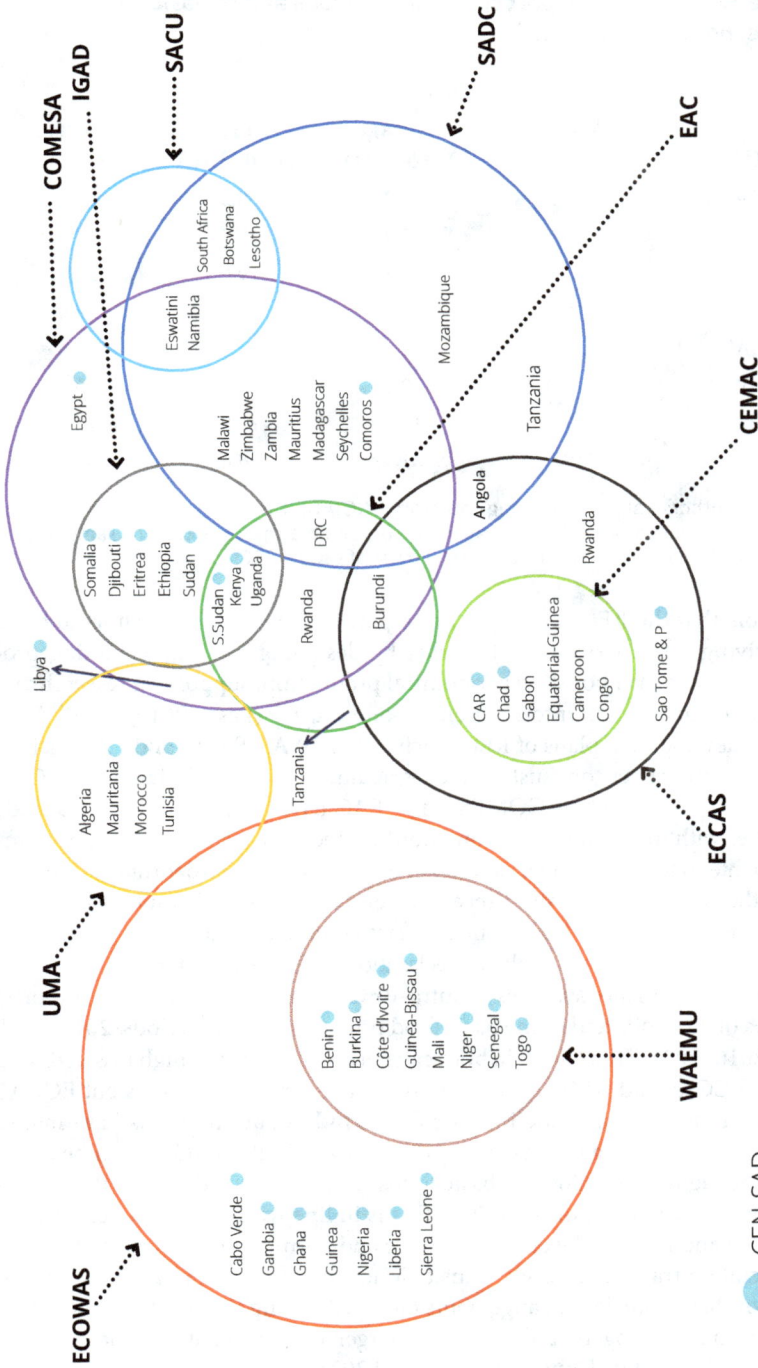

Source: MacLeod, Luke and Guepie (2023, p.41).
Notes: In 2024, Somalia became a member of EAC, while Burkina Faso, Mali and Niger announced that they were leaving ECOWAS and Sudan suspended its membership of IGAD. These changes are not shown in this graphic.

Table 5.1: Intra-REC export shares of all goods and of basic foods, by value, period averages (%)

| | All goods | | Basic food | |
REC	Average 2010–2015	Average 2016–2021	Average 2010–2015	Average 2016–2021
AMU	3.3	3.3	8.2	5.9
CEN-SAD	6.5	7.4	17.5	14.0
COMESA	8.3	10.3	19.6	20.4
EAC	18.2	18.0	23.4	31.6
ECCAS	2.3	2.5	29.6	30.6
ECOWAS	8.0	9.1	25.5	20.0
IGAD	10.9	19.6	12.9	19.7
SADC	19.1	21.4	38.6	33.3

Source: Authors' calculations using data from UNCTADSTAT.
Note: Since most African countries are members of more than one REC, the same trade flows will be counted in the intra-regional trade of several RECs.

the role that the RECs are expected to play in the continental effort to build a thriving intra-African food market by designing and implementing programmes aimed at boosting agricultural production, supporting research and catalysing investment into a regional food industry (Pasco 2019).

The development plans of RECs such as ECOWAS, SADC and EAC include regional initiatives on sustainable agriculture (MacLeod, Luke and Guepie 2023). RECs such as COMESA and EAC have pioneered simplified trade regimes with minimal customs and border checks to enable smallholder farmers to integrate into regional value chains through cross-border trade. In spite of this, the African Union's most recent review under CAADP's mutual accountability mechanism found that none of Africa's subregions was on track to meet its commitments under the Malabo Declaration (African Union n.d.).

Table 5.1 presents averages of intra-REC export (and conversely imports) shares of all goods and of the basic food products for two periods, 2010–2015 and 2016–2021. Trade flows in both categories are relatively high in SADC and EAC. ECCAS and AMU have the lowest trade flows in all goods but ECCAS performs strongly in trade in basic foods. Indeed, all the RECs have higher trade flows in basic goods than they do in all goods. It should further be noted that the higher trade flows in basic foods were maintained during the Covid-19 pandemic in most of the RECs. This is in line with the initiatives taken at country and regional levels to introduce and harmonise 'safe trade' measures to facilitate trade and goods transit (MacLeod and Guepie 2023). It is also known that 'groupage' arrangements emerged during the pandemic, whereby traders banded together to transport larger consignments of goods across borders (McCartan-Demie and MacLeod 2023).

Figure 5.6: Intra-African imports of basic food products (US$ million at 2022 prices), 2012–2022

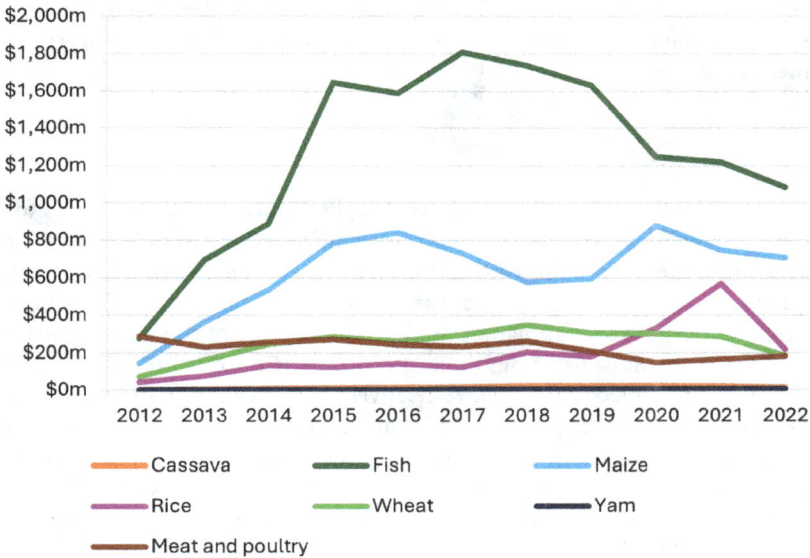

Source: Author's calculations based on United Nations (n.d.).
Notes: Adjustment to constant prices is based on World Bank (2023).

Cassava

Each of the basic foods tends to be traded within Africa in its own way. The vast majority of trade in cassava takes place in Eastern Africa,[2] with the sub-region accounting for 90 per cent of intra-African imports of that crop. This is somewhat surprising given that the continent's largest producers are in West and Central Africa, as shown in Chapter 3. This may reflect the fact these producers find it easier to export to North America and Europe, given their location on the Atlantic Ocean.[3] Indeed, in 2022, over 75 per cent of cassava exports by value from the continent's largest producers (Democratic Republic of the Congo, Ghana and Nigeria) went to the EU and North America, with only 11 per cent being exported to Africa.[4]

Rwanda alone imported around 80 per cent by value (but 53 per cent by volume) of total intra-African trade in cassava, while Tanzania was the largest intra-African exporter (with around 90 per cent of the total). Exports from the latter to the former accounted for 40 per cent of the total intra-African trade in cassava by value in 2022.[5]

Rwanda uses cassava primarily for food. Although Rwanda consumes far less cassava than many other African countries, as of 2021 the continent's main consumers tended to meet their needs through domestic production, importing only small amounts (whether from Africa or outside the continent) (author's analysis of FAOSTAT 2023b). Aside from Eastern Africa, there is also significant intra-African trade in cassava in several countries in Central, Western and Southern Africa, but very little in Northern Africa (which accounts for just 0.2 per cent of intra-African imports).[6] This is not surprising since cassava hardly features in the North African diet.

Fish

Intra-African trade in fish is more evenly split. Western Africa accounts for 47 per cent of intra-African fish imports (of which Côte d'Ivoire alone accounts for almost three-quarters), with Eastern Africa accounting for 14 per cent, Central Africa for 18 per cent, Southern Africa for 9 per cent and Northern Africa for 11 per cent (all figures are by value). Three of the four countries that import the most fish from the rest of Africa (Côte d'Ivoire, Algeria and South Africa) are coastal but have relatively high per capita incomes. Fish trade in these cases may be driven by an ability to afford a higher level of fish in the diet. Also, among the countries importing the most fish from the rest of Africa (by value) are Zambia (seventh), which is landlocked, and Democratic Republic of the Congo (fifth), which is almost landlocked (authors' analysis of UNCTAD 2023a).

Western Africa exports more fish than any other subregion on the continent (accounting for 54 per cent of exports), and around three-quarters of its intra-African fish imports come from within the region (authors' analysis of United Nations n.d. and UNCTAD 2023a). This is followed by Southern Africa (25 per cent), Northern Africa (14 per cent), Eastern Africa (5 per cent) and Central Africa (2 per cent). By far the largest intra-African exporters of fish are Mauritania, Namibia and Senegal, all of which are located on the Atlantic Ocean, which suggests strong demand for its fish varieties (UNCTAD 2023a). Despite this, most of the fish that Africa imports comes from outside the continent, and a substantial share of the continent's exports go to non-African countries. For example, Spain was the leading importer of African fish in 2022, primarily from its neighbour Morocco but also from other African countries (Chatham House 2021).

Maize

Intra-African trade in maize is concentrated in Eastern and Southern Africa, which account for 60 per cent and 29 per cent of intra-African imports, respectively. They also account for 49 per cent and 47 per cent of intra-African exports of this product, respectively. As these figures imply, the trade occurs between countries of Eastern and Southern Africa trading with one another; only a small share involves either imports from or exports to the other subregions (authors' analysis of United Nations n.d.).

As with cassava, some of the largest producers of maize (Nigeria and Egypt) are found in other subregions, even though Eastern and Southern Africa are the location of most of the continent's trade in this product. But, unlike the case of cassava, the apparent discrepancy (Egypt and Nigeria are leading producers but small players in the intra-African market) is not because these countries export their product outside Africa – it is because they use it for their domestic markets (authors' analysis of United Nations n.d. and FAOSTAT 2023b). Indeed, in 2021, Egypt exported only 2 per cent of its total maize production, while Nigeria exported less than 1 per cent (authors' analysis of FAOSTAT 2023b). Overall, Africa still has a deficit in maize (authors' analysis of FAOSTAT 2023b). As was seen in Chapter 2, most of the major partners for African countries' food imports and exports are outside the continent. Western Africa imports a substantial share of its maize from Latin America, while South Africa (which accounts for most of the continent's exports by itself) sends most of its maize exports to Asia and Italy (Chatham House 2021).

Meat and poultry

Intra-African trade in meat and poultry also occurs predominantly within and between Eastern and Southern Africa. In 2022, Southern Africa accounted for 44 per cent of intra-African imports and Eastern Africa accounted for 28 per cent by value. Southern Africa accounts for around two-thirds of intra-African exports of meat and poultry, with South Africa by itself exporting around half of the continent's meat and poultry by value. In neither of these two subregions does a single country dominate demand for imports. It is unclear whether this is due to higher volumes of meat being traded in Southern Africa, or higher values, as information on *volumes* of intra-African trade in meat and poultry is not available (authors' analysis of United Nations n.d.). Moreover, Southern Africa's exports are not predominantly driven by bovine meat or processed meat (which can be more expensive than other meats) (authors' analysis of UNCTAD 2023a).

Eastern Africa actually exports more meat by value than does Southern Africa, but around 80 per cent of its exports by value go to Western Asia, which may be explained by its geographical proximity and high demand from higher-income countries in that region (authors' analysis of UNCTAD 2023a). At the same time, most of Africa's meat imports are sourced from outside the continent (authors' analysis of Chatham House 2021). Africa has a deficit in both meat and poultry products (FAOSTAT 2023b).

Rice

Eastern Africa also dominates intra-African trade in rice. As of 2022, the subregion imported 56 per cent of total intra-African rice imports by value.

Almost nine-tenths of this rice came from within the subregion (rather from the rest of Africa). No single country dominates this share, though Uganda accounts for more than a third of the subregion's intra-African rice imports by itself. Central and Western Africa each accounted for 13 per cent of intra-African rice imports, while Southern Africa accounted for 16 per cent and Northern Africa less than 1 per cent. This is not because Eastern Africa imports the most rice. Western Africa imports over three times the level that Eastern Africa does, but it sources most of this rice from Asia, as do all of Africa's subregions (UNCTAD 2023a). Africa's exports of rice are overwhelmingly to other African countries, although some of these same countries also import rice from outside the continent, suggesting that some re-export but also that there could be potential to increase intra-African exports of rice at the expense of imports from the rest of the world (authors' analysis of Chatham House 2021).

Wheat

Intra-African imports of wheat are relatively evenly split between Southern Africa (31 per cent), Eastern Africa (22 per cent) and Northern and Western Africa (21 per cent each). Central Africa accounts for only 5 per cent of these imports. In Southern and Western Africa, most of these imports come from within the same subregion, but for the continent's other subregions this is not the case (though for Northern Africa significant trade flows may be missing from the UN Comtrade database for 2021 and 2022 as there is no data for Libya and Sudan, which are significant intra-African importers of wheat).[7] Central Africa's marginal role in intra-African trade may be partly explained by the fact that the region uses the least wheat (and products derived therefrom) than of any of the continent's other subregions. As of 2021, that subregion had the lowest domestic food supply of wheat and wheat products per capita basis of any subregion (author's analysis of FAOSTAT 2023b). Northern Africa's modest share in intra-African trade in wheat may seem surprising given that it produces more wheat and wheat products than any other region in Africa, accounting for 70 per cent of the continent's production in 2021. This could be because much of North Africa's wheat exports are sold outside the region, largely to Western Asia. Overall, most wheat that is traded in Africa comes from the rest of the world, even though African countries also export wheat outside the continent (Chatham House 2021). This suggests that reducing intra-African trade barriers could allow African producers to capture more business from the continental market (and costs of wheat may fall for producers and consumers if the costs of trading wheat within the continent can be brought below those of trading it with the rest of the world). Indeed, research published in 2022 suggests that reductions in intra-African trade costs have historically driven substantial increases in intra-African trade (Olney 2022). In 2022, African Development Bank launched an initiative aimed at boosting the continent's production of wheat, rice and other crops. This included plans to improve transport links between African countries, cutting the cost of trading (Ibukun 2022).

Yams

Intra-African imports in yams were dominated in 2022 by Southern Africa (accounting for around 73 per cent) and Western Africa (20 per cent).[8] Western Africa exports almost 98 per cent of the yams that are traded in the continent (and also accounts for an estimated 97 per cent of the continent's yam production) (authors' analysis of FAOSTAT 2023a; United Nations n.d.). This reflects the fact that Western Africa has much higher domestic food supply of yams per capita than any other subregion (75 kg per year, compared to 7 kg in central Africa and less than 1 kg in all other subregions) (FAOSTAT 2023b). Southern Africa accounts for a significant share of intra-African trade in yams. But intra-African trade in yams accounted for less than 0.02 per cent of the continent's production by volume in 2022 and reported world trade accounted for less than 0.2 per cent of the continent's production in 2021 (authors' analysis of FAOSTAT 2023a; United Nations n.d.). Yams produced in Africa are therefore overwhelmingly consumed within the country of production rather than being traded within the continent or internationally. Yam imports into Southern Africa may be linked to the crop's use for traditional healing in that subregion (Beinart 2020).

For many of these products, aside from fish, intra-African trade occurs between countries of the same subregion, often Eastern and Southern Africa and to a lesser extent Western Africa. This may reflect the fact that there are functioning regional free trade areas within these subregions. The fact that Eastern and Southern Africa dominate trade in most of these products (ahead of Western Africa) suggests that these RECs may be more effective in promoting intra-regional trade in these foods. This aligns with research findings vis-à-vis general trade (Kassa and Sawadogo 2021).

5.3 Informal cross-border food trade

Informal cross-border trade (ICBT) is ubiquitous in Africa but defining it has proved to be elusive. The term 'informal' often evokes an allusion to illegal activities. In practice, however, informal cross-border traders use both formal and informal routes. In the latter case, the intent may not necessarily be to evade customs control or border taxes but rather to avoid cumbersome border procedures, especially when the value of the consignment is small. High and arbitrary charges levied at borders, social marginalisation from formalisation efforts and closure of official borders are some reasons why cross-border traders use informal routes (Nakayama, 2022; Nkendah, 2020; Wiseman, 2022).

As previously noted, some RECs have adopted simplified trade regimes for small consignments.

Estimates of the extent of ICBT vary according to the methodology that is used but suggest that it accounts for a significant proportion of intra-African trade (Bouët, Pace and Glauber 2018; Walkenhorst 2020; Gaarder, Luke and Sommer 2021). Gaarder, Luke and Sommer (2021) suggest that the average value of ICBT lies between US$10.4 billion and US$24.9 billion, representing 7–16 per cent of total intra-African trade or 30–72 per cent of formal trade between neighbouring countries. This is comparable with estimates for SADC and for COMESA, where ICBT was assessed to be up to 40 per cent of recorded intra-REC trade (Afrika and Ajumbo 2012; Nshimbi and Moyo 2017).

The Famine Early Warning Systems Network (FEWS NET), an initiative established by the US Agency for International Development, collects data on informal cross-border trade at selected border posts across Eastern, Northern and Southern Africa.[9] Analysing this data for the eight basic food products discussed in this book suggests that informal trade (by volume) could account for a significant share of total trade, depending on the product, as shown in Table 5.2 below.

In relation to ICBT food composition, the World Bank (Walkenhorst 2020) found that for Uganda and Rwanda nine of the top 10 ICBT products are food products. Engel and Jouanjean (2013, p.13) found cereals, tubers like cassava and yam, fruits and vegetables, and livestock products to be widely traded in West Africa.

Table 5.3 shows that, for several food products, small-scale cross-border trade is the main channel through which these goods are imported and exported. These figures may be higher for Uganda and Rwanda than for some other countries since both are landlocked, eliminating seaborne trade, which is more likely to be formal as it must pass through a port.

With food products accounting for the largest share of ICBT, it remains an important source of affordable food for many households, not only in rural

Table 5.2: Lower-bound estimated shares of informal cross-border trade in total trade between 14 countries in Eastern, Northern and Southern Africa for selected food products, 2022

	Food product					
	Cassava and derivative products	Fish	Maize and derivative products	Rice and derivative products	Wheat and derivative products	Yams
Share of informal trade	4%	48%	21%	18%	1%	1%

Source: Authors' calculations based on Chatham House (2021); FEWS NET Famine Early Warning Systems Network (n.d.); United Nations (n.d.).

Table 5.3: Shares of small-scale cross-border trade in total trade, food products where the former accounts for at least 50 per cent of the latter, 2017

Uganda imports	Uganda exports	Rwanda imports	Rwanda exports
Bananas, 100%	Dried fish, 98%	Preserved fish, 100%	Swine meat, 100%
Wheat flour, 91%	Live bovine animals, 92%	Coffee, 99%	Bovine meat, 100%
Cassava, 89%		Bananas, 94%	Live poultry, 100%
Vegetable oil, 89%		Beer, 86%	Dried fish, 97%
Fruit juice, 85%		Potatoes, 55%	Milk and cream, 94%
Dried legumes, 69%		Dried leguminous. vegetables, 54%	Prepared fish, 92%
Leguminous vegetables, 69%		Cereal flour, 50%	Swine meat, 100%
Dried fish, 66%			Sugar, 64%
Onion and garlic, 60%			Cereal flour, 64%

Source: Adapted from Walkenhorst (2020, p.12).

areas (Zarrilli and Linoci 2020) but also across African cities (Skinner and Watson 2020, p.127).

Summary

The role of intra-African trade in meeting basic food supply varies from commodity to commodity. In some cases (fish), it may perform well, playing an indispensable role in transferring food from countries that are sizeable exporters, perhaps due to geographical advantages, to others that may not be able to meet domestic demand, either due to being landlocked or having higher-income economies where higher wages and stronger currencies may mean that it is more affordable to import fish from other countries across the continent. For commodities like yams and for some of Africa's largest producers of maize, intra-African trade and indeed international trade may be of marginal importance, and needs are met through domestic supply. This appears to be true even when data on informal trade is considered. While African demand for cassava is largely met through domestic production, Africa's supply of this crop exceeds its demand, and the balance is exported outside the continent (author's analysis of FAOSTAT 2023b; Silva et al. 2023; Yuan et al. 2024). For wheat, maize, meat and poultry, substantial shares of Africa's imports come from outside the continent, while at the same time exports go in the other direction, suggesting that a reduction in costs of intra-African trade could allow producers to capture greater market share. Even with fish, where intra-African trade has an important role, Africa imports more from the rest of the world than within the continent, while at the same time exporting much of its catch to non-African countries, although it is not known how much of what is recorded as African fish exports is via illegal fishing (see Chapter 9).

In addition to reducing intra-African trade costs and boosting production, a shift towards crops in which Africa has high potential could also help the continent to close its food trade deficit. For example, cassava is one of the most productive crops in the world (Danino 2023). African countries have the opportunity to significantly increase their production of yams, another crop in which the continent has (almost) no trade deficit (Owusu Danquah et al. 2022; FAOSTAT 2023b). Aquaculture has the potential to reduce Africa's fish imports (Eyayu, Getahun and Keyombe 2023; Ragasa et al. 2022), along with better controls over coastal and offshore fishing.

Notes

[1]   Based on averages for the period 2017–2021. Data is from UNCTADSTAT.

[2]   In this chapter, we use regional and subregional classifications following the UN Statistics Division's classification (UNSTATS n.d.).

[3]   Interestingly, the Organisation for Economic Co-operation and Development (OECD) estimates that the costs of insurance and freight for Ghana, Nigeria and Democratic Republic of the Congo of trading cassava and other similar products with the United States and Europe are similar to the cost of trading these products within Africa. This may suggest that there are other reasons for these producers to prefer the European and North American markets. These could be the difficulty in using trade preferences within Africa (Author's analysis based on OECD n.d.). On the challenges of using preferences, see UNCTAD (2023b).

[4]   Authors' analysis of United Nations (n.d.). Accessed via World Bank World Integrated Trade Solution.

[5]   Authors' analysis of United Nations (n.d.). Accessed via World Bank World Integrated Trade Solution.

[6]   Authors' analysis of United Nations (n.d.). Accessed via World Bank World Integrated Trade Solution.

[7]   While the overall picture of intra-African trade in wheat being dominated by Eastern, Northern, Southern and Western Africa had been the case already in 2021, the specific shares changed and Intra-African trade in wheat and wheat products declined around one-third from 2021 to 2022. The fact that fewer countries reported trade data for 2022 accounts only for three percentage points of this decline. A decline in Africa's total

wheat production accounts for around 10 percentage points. A possible explanation for the remaining decline is that Africa's wheat exports to the rest of the world increased as other countries sought wheat from alternative sources as Russia's invasion of Ukraine disrupted supplies from those countries, meaning that there was less wheat to trade within the continent. For example, this appears to have affected Sudan's wheat exports but political instability in that country could also be part of the explanation (Authors' analysis of United Nations n.d.; Chatham House 2021).

[8] In 2021, Central Africa accounted for 7 per cent of intra-African yam imports. Data for Cameroon and Gabon, the two countries in Central Africa with the highest intra-African yam imports in 2021, was missing in 2022, so these calculations add in figures for these two countries based on 2021 import levels.

[9] The border posts are in Djibouti, Ethiopia, Kenya, Malawi, Mozambique, Rwanda, Somalia, South Africa, South Sudan, Sudan, United Republic of Tanzania, Uganda, Zambia and Zimbabwe.

References

African Union (n.d.) *3rd CAADP Biennial Review Report*, Addis Ababa: African Union. https://perma.cc/5NAA-PFMV

Afrika, Jean-Guy; and Ajumbo, Gerald (2012) 'Informal Cross Border Trade in Africa: Implications and Policy Recommendations', *Africa Economic Brief*, vol. 3, no. 10, pp.1–13.

Beinart, William (2020) 'South African Yams Are a Miracle Drug. Can They Be Saved?', *The National Interest*, 12 February. https://perma.cc/979T-CPEP

Bouët, Antoine; Pace, Kathryn; and Glauber, Joseph W. (2018) Informal Cross-Border Trade in Africa: How Much? Why? And What Impact?, IFPRI Discussion Paper 01783, International Food Policy Research Institute. https://perma.cc/NDE9-A5WA

Chatham House (2021) 'resourcetrade.earth', ResourceTrade.Earth. https://perma.cc/FFU2-5QYM

Cust, James; and Zeufack, Albert (2023) *Africa's Resource Future: Harnessing Natural Resources for Economic Transformation during the Low-Carbon Transition*, World Bank Publications.

Danino, Daniel (2023) 'Council Post: How Cassava Could Impact the Future of Agriculture in Africa', *Forbes*. https://perma.cc/3FLV-S7FP

Engel, Jakob; and Jouanjean, Marie-Agnès (2013) 'Barriers to Trade in Food Staples in West Africa: An Analytical Review', ODI. https://perma.cc/Z633-KA6X

Eyayu, Alamrew; Getahun, Abebe; and Keyombe, James L. (2023) 'A Review of the Production Status, Constraints, and Opportunities in East African Freshwater Capture and Culture Fisheries', *Aquaculture International*, vol. 31, no. 4, pp.2057–78. https://doi.org/10.1007/s10499-023-01071-1

FAOSTAT (2023a) 'Crops and Livestock Products', FAOSTAT. https://perma.cc/P5R7-AV3V

FAOSTAT (2023b) 'Food Balances (2010-)', FAO. https://perma.cc/H44G-48E2

FEWS NET Famine Early Warning Systems Network (n.d.) 'Markets and Trade'. https://perma.cc/VJA2-E5BN

Gaarder, Edwin; Luke, David; and Sommer, Lily (2021) *Towards an Estimate of Informal Cross-Border Trade in Africa*. Addis Ababa: Economic Commission for Africa. https://perma.cc/J5RL-6P7T

Ibukun, Yinka (2022) 'Lender Has $1 billion Plan to Wean Africa Off Russian Wheat', African Development Bank Group. https://www.afdb.org/en/news-and-events/lender-has-1-billion-plan-wean -africa-russian-wheat-50087

International Monetary Fund (1992) 'Global Price Index of All Commodities', FRED, Federal Reserve Bank of St. Louis. https://perma.cc/9NJC-R42G

Kassa, Woubet; and Sawadogo, Pegdéwendé N. (2021) 'Trade Creation and Trade Diversion in African Recs: Drawing Lessons for AfCFTA', The World Bank (Policy Research Working Papers, 9761). https://doi.org/10.1596/1813-9450-9761

MacLeod, Jamie; and Guepie, Geoffroy (2023) 'How the Covid-19 Crisis Affected Formal Trade', in Luke, David (ed.) *How Africa Trades*, London: LSE Press. https://doi.org/10.31389/lsepress.hat

MacLeod, Jamie; Luke, David; and Guepie, Geoffroy (2023) 'The AfCFTA and Regional Trade', in Luke, David (ed.) *How Africa Trades*. London: LSE Press. https://doi.org/10.31389/lsepress.hat

McCartan-Demie, Kulani; and MacLeod, Jamie (2023) 'How the Covid-19 Crisis Affected Informal and Digital Trade', in Luke, D. (ed.) *How Africa Trades*, London: LSE Press. https://doi.org/10.31389/lsepress.hat

Nakayama, Yumi (2022) 'Why Do Informal Cross Border Traders (ICBTs) Operate Informally? The Paradox of the Formalization of ICBTs in Africa', *ASC-TUFS Working Papers*, 2. https://perma.cc/8YTX-MYKF

Nkendah, Robert (2010) 'The Informal Cross-Border Trade of agricultural commodities between Cameroon and its CEMAC's Neighbours', in. NSF/ AERC/IGC Conference, Mombasa. https://perma.cc/TKL3-XZQ2

Nshimbi, Christopher; and Moyo, Inocent (2017) *Migration, Cross-Border Trade and Development in Africa*, Springer. https://link.springer.com/book/10.1007/978-3-319-55399-3

OECD (n.d.) 'International Transport and Insurance Costs of Merchandise Trade (ITIC)', OECD.Stat. http://stats.oecd.org/Index.aspx?DataSetCode=CIF_FOB_ITIC#

Olney, William W. (2022) 'Intra-African trade', *Review of World Economics*, vol. 158, no. 1, pp.25–51. https://doi.org/10.1007/s10290-021-00421-6

Owusu Danquah, Eric; Danquah, Frank Osei; Frimpong, Felix; Obeng Dankwa, Kwame; Kumari Weebadde, Cholani; Ennin, Stella Ama; Asante, Mary Otiwaa Osei; and Badu Brempong, Mavis et al. (2022) 'Sustainable Intensification and Climate-Smart Yam Production for Improved Food Security in West Africa: A Review', *Frontiers in Agronomy*, vol. 4. https://doi.org/10.3389/fagro.2022.858114

Pasco, Allan H. (2019) 'Feeding Africa through Increased Intra-African Food Trade', *Annales des Mines – Réalités industrielles*, Août 2019, no. 3, pp.72–75. https://doi.org/10.3917/rindu1.193.0072

Ragasa, Catherine; Charo-Karisa, Harrison; Rurangwa, Eugene; Tran, Nhuong; and Mashisia Shikuku, Kelvin (2022) 'Sustainable Aquaculture Development in Sub-Saharan Africa', *Nature Food*, vol. 3, no. 2, pp.92–94. https://doi.org/10.1038/s43016-022-00467-1

Rettig, Michael; Kamau, Anne W.; and Muluvi, A. S. (2023) 'The African Union Can Do More to Support Regional Integration', Brookings. https://perma.cc/AD9L-L9DQ

Silva, João V.; Jaleta, Moti; Tesfaye, Kindie; Abeyo, Bekele; Devkota, Mina; Frija, Aymen; Habarurema, Innocent; and Tembo, Batiseba et al. (2023) 'Pathways to Wheat Self-Sufficiency in Africa', *Global Food Security*, vol. 37, p.100684. https://doi.org/10.1016/j.gfs.2023.100684

Skinner, Caroline; and Watson, Vanessa (2020) 'The Informal Economy in Urban Africa: Challenging Planning Theory and Praxis', in *The Informal Economy Revisited*, London: Routledge, pp.123–31. https://www.taylorfrancis.com/chapters/oa-edit/10.4324/9780429200724-21/informal-economy-urban-africa-caroline-skinner-vanessa-watson

UNCTAD (2023a) 'Merchandise Trade Matrix, Annual UNCTAD, Stat'. https://unctadstat.unctad.org/datacentre/dataviewer/shared-report/16a48b97-54fb-4349-8972-2ba3dcc09a97

UNCTAD (2023b) 'Simpler Rules of Origin Needed to Boost Free Trade in Africa, Study Shows UNCTAD, Prosperity for All'. https://perma.cc/G7YX-SYPV

United Nations (n.d.) 'UN Comtrade Database'. comtrade.un.org. https://perma.cc/5L26-AT6H

United Nations Economic Commission for Africa, AU, AfDB and UNCTAD (2019) 'Assessing Regional Integration in Africa IX', Addis Ababa: Economic Commission for Africa (Assessing Regional Integration in Africa).

UNSTATS (n.d.) 'Statistics Division Methodology Standard Country or Area Codes for Statistical Use (M49)', United Nations. https://perma.cc/7GPU-US7E

Walkenhorst, Peter (2020) 'Monitoring Small-Scale Cross-Border Trade in Africa: Issues, Approaches, & Lessons', International Bank for Reconstruction and Development/The World Bank. https://perma.cc/C9Y7-3EPF

World Bank (2023) 'GDP Deflator (Base Year Varies by Country)', The World Bank, Data. https://perma.cc/96MV-UK5F

Wiseman, Eleanor (2022) 'Trade, informality, and corruption: Evidence from small-scale traders in Kenya', International Growth Centre, 2 February. https://perma.cc/DS2P-VB85.

Yuan, Shen; Saito, Kazuki; van Oort, Pepijn A. J.; van Ittersum, Martin K.; Peng, Shaobing; and Grassini, Patricio (2024) 'Intensifying Rice Production to Reduce Imports and Land Conversion in Africa', Nature Communications, vol. 15, no. 1, p.835. https://doi.org/10.1038/s41467-024-44950-8

Zarrilli, Simonetta; and Linoci, Mariangela (2020) What Future for Women Small-Scale and Informal Cross-Border Traders when Borders Close?, UN Trade & Development UNCTAD. https://perma.cc/K2NN-C3T3

6. Expected impact of the African Continental Free Trade Area on food security

Jamie MacLeod

Intra-African trade is a highly important component of Africa's agricultural trade. Much of this trade comprises goods such as vegetables, fish, vegetable oils, fruits and diary (which tend to have greater earning potential than other agricultural or food products), as detailed in Chapter 5. Intra-African trade in agricultural machinery and fertilisers is also significant while that in staple foods, such as millet, sorghum and rice, is relatively limited according to official trade flows, though likely higher once small-scale informal trade is taken into account. The effort to establish a continent-wide free trade area in the form of the African Continental Free Trade Area (AfCFTA) is driven by the recognition that a liberalised trade regime across the continent could drive further growth in intra-African trade including informal trade formalisation as tariffs not already covered by regional trade agreements and non-tariff barriers fall.

This is the background against which this chapter estimates the potential impact of the AfCFTA on the agriculture sector by means of a detailed partial equilibrium model using recently available tariff schedules for African countries. The impact is forecast to be relatively small over the short-run timespan that is covered by the modelling approach. Intra-African trade in the sector as a whole is expected to increase by 5.4 per cent, equivalent to $1,015 million annually, in a scenario of full tariff liberalisation under the AfCFTA. These results are modest, but consistent across all tests of the model to parameter sensitivity analyses, and reflect a scenario in which all trade is liberalised, without recourse to product exclusions (in reality some of this trade may be excluded from liberalisation and the impact of the AfCFTA may be limited further).

How to cite this book chapter:

MacLeod, Jamie (2025) 'Expected impact of the African Continental Free Trade Area on food security', in: Luke, David (ed) *How Africa Eats: Trade, Food Security and Climate Risks,* London: LSE Press, pp. 125–157. https://doi.org/10.31389/lsepress.hae.f License: CC-BY-NC 4.0

Why are the results modest? The technical answer is that trade modelling, such as that deployed in this chapter, relies on amplifying existing trade and there is simply not that much existing formal trade between African countries. Intra-African suppliers account for just 16 per cent of Africa's imports of agricultural and food products. Most of that existing trade already flows through Africa's pre-existing free trade arrangements of the regional economic communities (RECs), so the tariffs faced by that trade are also already low, averaging just 2.9 per cent (on an import-weighted basis).

Nevertheless, where the AfCFTA will have most impact, in the immediate term, is on trade in relatively higher unit-value products including fish and seafood, vegetables, preparations of cereals, vegetable oils, fruits and dairy. There are also relatively sizeable opportunities for exports of sugar and coffee. In the upstream part of the value chain, there are important opportunities for exporters of agricultural machinery and fertilisers.

What these results suggest is that, if the AfCFTA is going to substantially boost the agriculture sector and food security in Africa, it needs to go far beyond merely reducing tariffs. This aligns with the results of other modelling assessments, such as those by the World Bank (2020), the IMF (2019), and the ECA (2021). While less detailed or focused on agriculture specifically, those complementary assessments suggest that much of the impact of the AfCFTA will arise only if it can be effectively used as a tool for reducing non-tariff barriers and stimulating investment. Such non-tariff barriers are found to be higher in the agriculture sector than in other sectors on average (UNECA et al. 2019). Yet, unlike reducing tariff barriers, these benefits of the AfCFTA are not automatic and will require much more work to unpeel the layers of non-tariff challenges facing African traders. Chapter 7 identifies and discusses pathways for the AfCFTA to discipline non-tariff barriers, one aspect of this work.

6.1 Assessing the impact of the African Continental Free Trade Area on the agriculture sector

The analysis uses a partial equilibrium modelling approach (a full elaboration of the model is included in Appendix A). Partial equilibrium models specialise in providing detail on one part or sector of the economy over the short to medium term. They can incorporate the latest tariff schedules and trade flows and provide detailed analysis, which makes them suitable for assessing product-specific impacts within the agriculture sector. This also allows the analysis to use the most recently available tariff schedules for African countries and recent trade data. It should nevertheless be considered as best reflecting *short-run* first-order effects, after which general equilibrium effects are likely to be increasingly important.

Examples of partial equilibrium models in the context of the AfCFTA include those by Mulugeta (2020) on Ethiopia, Bayale, Ibrahim and

Atta-Mensah (2022) on Ghana, Fouda Ekobena et al. (2021) on Central Africa, Oyelami (2021) on Nigeria, Seti and Daw (2022) on the South African agricultural sector, Lunenborg and Roberts (2021) on the Economic Community of West African States (ECOWAS) region and Ossadzifo Wonyra and Bayale (2022) on Togo. Like those assessments, this one focuses on the short run and does not account for broader economy-wide linkages that would tend to be more important over a longer time horizon, including linkages between factor incomes and expenditures and broader macroeconomic adjustments, such as changes to the exchange rate, investment rates, or constraints within endowment markets.

The model simulates the response of imports and other variables to changes in the tariff rate. The underlying model assumes imperfect substitution between different import sources (what is known as the Armington assumption). Goods imported from different countries, although similar, are imperfect substitutes (maize from Kenya is an imperfect substitute for maize from Uganda). Within this assumption the representative consumer determines the level of imports of a good through a two-stage process. First, given an import price index, they choose the level of total spending on a 'composite import good' (say, imported maize). The relationship between changes in the import price index and the impact on total imports is determined by a given 'demand elasticity'. Then, within this composite good, they allocate spending among the different sources of the good, depending on the relative price of each source (say, choose more rice from Tanzania and less from Egypt). The extent of the between-source allocative response to a change in the relative price is determined by the 'substitution elasticity'. A full specification of the model is given in Appendix A.

The model is designed to reduce tariffs only on intra-African trade that is not covered by pre-existing free trade arrangements, such as those of the RECs. In other words, Article 19 of the AfCFTA, Conflict and Inconsistency with Regional Arrangements of the Protocol on Trade in Goods, is fully applied, meaning that trade under pre-existing intra-African preferential trade agreements will continue to be governed by those pre-existing arrangements.[1]

6.2 Structure of existing African trade and tariffs

The results of the model are driven primarily by the shape and form of pre-existing tariffs and trade flows. Table 6.1 shows what these are.[2]

To understand the eventual results of the model, it is important to appreciate a differentiation between two types of intra-African trade. The first is intra-African trade in its entirely. This involves both trade *within* pre-existing regional trade arrangements, such as that between Kenya and Uganda within the East African Community (EAC), as well as trade *between* regional arrangements, such as from Kenya in the EAC to Ghana in ECOWAS. The second is the subset of intra-African trade that is not already covered by pre-existing regional trade agreements. This would concern only the latter, that is, to use our example, trade between Kenya in the EAC and Ghana in ECOWAS.

If we consider intra-African trade *in its entirety*, the average imported-weighted tariffs on intra-African trade are relatively low, at 1.9 per cent, compared to 9 per cent faced by imports from outside the continent. This is because the majority of intra-African trade already flows through pre-existing regional trade arrangements. Kenya (currently) trades much more with its East African neighbours than with African countries further afield, like Ghana.

If we consider instead only intra-African trade that is not already covered by regional trade agreements, tariffs are much higher. The average import-weighted tariff faced by intra-African exports *outside of regional trade agreements* is considerably higher, at 18 per cent for vegetable products and 19 per cent for foodstuffs. These tariffs are substantial and it is this trade that is scheduled to be liberalised by the AfCFTA and which will drive the results of the modelling.

The intra-African tariffs for the Harmonised System (HS) sector headings associated with agriculture are among the highest of the different sector headings. Tariffs on intra-African trade of 'foodstuffs' are the highest.

Those on vegetable products are fourth highest, behind the textiles and apparel and miscellaneous sections. Nevertheless, tariffs are not especially high, on an import-weighted basis, for any of this trade.

Table 6.1 also shows existing trade flows. These are important for the results of the modelling because what such models do is to effectively scale up (or down) existing trade flows, in line with the impact that the model estimates that tariff reforms will have on those flows. The more pre-existing trade there is, the more there is for the model to scale.

What the model cannot do is predict the creation of wholly new trade flows. In the 2017–2019 reference period for the model, total intra-African trade averaged $87 billion a year, while total imports from external suppliers outside the continent amounted to $513 billion. This puts intra-African trade as a share of total African imports at around 14.5 per cent, consistent with other estimates (ECA 2021). Intra-African trade flows in the foodstuffs section amounted to $6.8 billion, while those of vegetable products amounted to $5.2 billion. This means that the intra-African share of trade in these sections is 25 per cent and 11 per cent, respectively. To put that into context, intra-African trade is stronger than for the average across all trade (which was 14.5 per cent) in foodstuffs, and weaker in trade in vegetable products. In the context of the model, this means that there is both ample demand that could be met and replaced by intra-African suppliers but also that scaling up production to meet this demand could prove more difficult for vegetable products.

Table 6.2 narrows down the focus to the different parts of the agriculture value chain (as introduced in Chapters 2 and 3). As a reminder, foods include products like grains, tubers, meats and fish. Agricultural raw materials include cocoa beans, cotton, coffee, spices, wood and rubber. Agricultural inputs are mostly fertilisers and herbicides. Agricultural capital goods

Table 6.1: Tariffs and existing trade flows, by HS section, 2017–2019

	Simple average tariffs (%)	Average import-weighted tariffs (%)[†]		Existing trade flows (US$bn)	
		Intra-African*	External	Intra-African	External
Animal and animal products	16.7	2.8	13	3.1	13
Vegetable products	14.7	3.1	8	5.2	41
Foodstuffs	25.1	3.8	22	6.8	21
Mineral products	5.4	0.8	2	23.6	71
Chemicals and allied industries	5.6	1.8	6	7.7	47
Plastics/rubbers	10.2	2.6	10	2.8	27
Raw hides, skins, leather and furs	13.1	1.9	22	0.1	2
Wood and wood products	11.1	2.2	10	2.4	16
Textiles and apparel	17.3	3.2	19	2.7	30
Footwear/headgear	20.0	2.9	20	0.5	4
Stone/glass	14.2	1.4	15	7.3	10
Metals	10.4	1.4	11	10.8	42
Machinery/electrical	7.0	2.3	7	6.2	109
Transportation	8.3	2.3	12	6.4	60
Miscellaneous	11.9	3.6	11	1.3	19
TOTAL	**11.6**	**1.9**	**9**	**87**	**513**

Source: Author, based on the Centre d'études prospectives et d'informations internationales's (CEPII) Base pour l'Analyse du Commerce International (Database for the analysis of international trade) (BACI) trade dataset and tariffs reported in the World Trade Organization (WTO) Integrated Database and International Trade Centre (ITC) MAcMap database for the latest available years.

Notes: * If intra-African tariffs seem unusually low, it is because it includes *intra-REC* free trade area (FTA) trade which for the purpose of the model is simulated as being zero, i.e. it will be unaffected by the AfCFTA in accordance with Article 19.

[†] Average import-weighting is used here as an intuitive and transparent aggregation method but, being endogenous, should not be interpreted as indicative of effective protection. This does not affect the modelling, which is conducted at the HS6 level of disaggregation.

are tractors, agricultural machinery (such as seeders, harvesters and dryers) and agricultural tools.

The agriculture sector faces higher tariffs than other HS sections on average. The average import-weighted tariff on intra-African imports (in their entirety, including both those within and outside regional trade agreements)

Table 6.2: Tariffs and existing trade flows, by segment of the agricultural sector, 2017–2019

	Simple average tariffs (%)	Average import-weighted tariffs (%)		Existing trade flows (US$bn)	
		Intra-African	External	Intra-African	External
All food	17.9	3.3	13	13.5	71
Agricultural raw materials	9.8	3.4	9	2.2	9
Agricultural capital	4.9	1.5	6	1.0	10
Agricultural inputs	2.3	1.0	3	2.1	5
TOTAL		2.9	11	19	96

Source: Author, based on the CEPII-BACI trade dataset and tariffs reported in the WTO Integrated Database and ITC MAcMap database for the latest available years.

within the agriculture sector is 2.9 per cent (compared to 1.9 per cent for all imports). As would be expected, given that most tariff schedules aim to support productivity, average tariffs are lower on agricultural inputs and capital goods and higher on final consumption goods.

Data on existing trade flows shows that total African import demand is largest for foods. Intra-African imports of foods are 11 times that of intra-African imports of agricultural raw materials. This is the part of the value chain where most value currently exists and where, in the modelling results, we will expect to see the largest nominal potential value in the AfCFTA.

Africa is highly reliant upon capital imported from outside the continent. Intra-African suppliers account for only 9 per cent of all imports of agricultural capital (such as tractors, agricultural machinery and tools) imported by African countries (Figure 6.1). In contrast, 28 per cent of import demand for agricultural inputs, 18 per cent of import demand for agricultural raw materials and 16 per cent of import demand for food is met with intra-African suppliers. By comparing these shares with the average intra-African share in total trade, which as mentioned above was 14.5 per cent, we determine that African countries trade more among themselves in agricultural goods than in other products.

Table 6.3 provides further disaggregation by the main products traded under each part of the agriculture value chain. Cells are shaded in green or blue according to their relative values.

The AfCFTA will have the greatest effect where it will be reducing high tariffs on intra-African trade, where there exists some intra-African trade to scale up, and where there is ample external trade to substitute away from. We can already identify where the AfCFTA is likely to have most impact. This

Figure 6.1: Africa's agricultural trade by source (US$ billion, current prices), 2017–2019

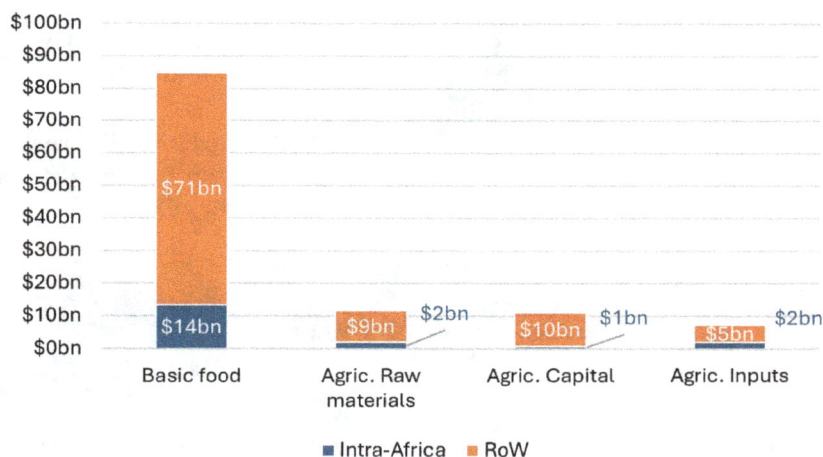

Source: Author, based on the CEPII-BACI trade dataset and tariffs reported in the WTO Integrated Database and ITC MAcMap database for the latest available years.

includes products with relatively high existing intra-African tariffs *and* existing intra-African trade flows. Prime examples are fish and seafood, vegetables, sugar, coffee and fruits.

We can also identify where intra-African trade is unlikely to be substantially affected by the tariff reductions under the AfCFTA. This is the case with products where either tariffs on intra-African trade are already very low or existing intra-African trade is too small to be substantially scaled up. For instance, wheat is one of the most important imports yet intra-African trade is very small, despite low prevailing intra-African tariffs. Tariff reductions under the AfCFTA are unlikely, in themselves, to remedy such circumstances. Similarly, tariffs are already low (on average) on intra-African trade in millet, soya beans, animal food and fodder, tea, agricultural capital and inputs.[3]

The key points of this section are threefold. First, average import-weighted tariffs on intra-African trade are low because most of this trade is already covered by Africa's pre-existing regional free trade arrangements (such as those of the EAC or ECOWAS). Second, within the agriculture value chain section, tariffs are highest on final consumption goods, including fish and seafood, vegetables, sugar, coffee and fruits. We will expect the AfCFTA to have the strongest impact on these products. Third, Africa is also, in general, highly dependent on imports from outside the continent of food security crops like wheat and maize. Low prevailing tariffs on these products suggest that tariff reductions under the AfCFTA will have little impact, however, on boosting intra-African trade in these products.

Table 6.3: Tariffs and existing trade flows, by segment of the agricultural sector, 2017–2019

		Average import-weighted tariffs (%)		Existing trade flows (US$bn)	
		Intra-African	External	Intra-African	External
	Vegetables	2.9	17	0.7	3
	Wheat	2.6	6	0.4	15
	Beef	1.3	13	0.2	2
	Dairy	1.7	9	0.5	4
	Fish and seafood	6.4	12	1.9	4
	Fruits	6.6	27	0.8	1
	Maize	3.1	4	0.5	4
	Millet	1.0	9	0.01	0.04
	Palm oil	1.0	14	0.6	5
	Preparations of cereals	3.4	15	0.7	3
All food	Sorghum	0.9	7	0.1	0.1
	Soya beans	0.1	2	0.03	1
	Vegetable oils	2.5	9	0.5	3
	Beverages	2.6	63	0.9	2
	Sugar	3.1	17	1.4	5
	Citrus fruit	6.8	20	0.1	0
	Tobacco	4.3	19	1.0	2
	Nuts	5.9	21	0.0	0
	Poultry	1.7	25	0.2	2
	Rice	0.2	8	0.3	6
	Other food	1.8	16	2.7	9
	All foods	**3.3**	**13**	**13.5**	**71**
	Coffee	10.0	13	0.3	1
	Tea	0.7	9	0.4	1
Agricultural raw materials	Cocoa	5.9	18	0.2	1
	Wood	1.1	4	0.4	3
	Flowers	6.6	24	0	0
	Fibres	4.2	12	0.2	3
	Cotton	0.2	1	0.2	0

(Continued)

Table 6.3: (Continued)

		Average import-weighted tariffs (%)		Existing trade flows (US$bn)	
		Intra-African	External	Intra-African	External
Agricultural raw materials	Other agricultural raw materials	3.4	7	0.5	2
	All agricultural raw materials	**3.4**	**9**	**2.2**	**9**
Agricultural capital	Machinery	1.8	5	0.7	7
	Tools	0.3	12	0.01	0.1
	Tractors	0.5	9	0.2	3
	All agricultural capital	**1.5**	**6**	**1.0**	**10**
Agricultural inputs	Fertilisers	1.0	2	1.8	3
	Insecticides	1.3	5	0.3	2
	All agricultural inputs	**1.0**	**3**	**2.1**	**5**
	TOTAL	**2.9**	**11**	**19**	**96**

Source: Author, based on the CEPII-BACI trade dataset and tariffs reported in the WTO Integrated Database and ITC MAcMap database for the latest available years.
Notes: Table is shaded with darker cells showing larger values. Green denotes cells relating to intra-African trade and blue denotes cells relating to imports from outside the continent.

6.3 Impact of the African Continental Free Trade Area

Aggregate results are presented in Table 6.4 for all HS section headings. The AfCFTA is estimated to have the potential to boost total intra-African trade by 5.7 per cent, equivalent to almost $5 billion, in the short term. All exports within the continent must by definition be equal to all imports from within the continent; therefore, intra-African trade here can be considered equivalently to exports or imports.

The relatively low impact of the AfCFTA from tariff elimination alone is consistent with estimates from other partial equilibrium models (Bayale, Ibrahim and Atta-Mensah 2022; Fouda Ekobena et al. 2021; Lunenborg and Roberts 2021; Mulugeta 2020; Oyelami 2021; Seti and Daw 2022). These focus on the shorter to medium term and tend to forecast somewhat lower magnitudes of impact than general equilibrium models. Assumptions on the impact of the AfCFTA on non-tariff barrier reductions and trade facilitation improvements account for much of the much larger estimated effects of the AfCFTA in the

Table 6.4: Impact of the AfCFTA, by HS section

	AfCFTA impact				
	Increase in intra-African trade		Trade diversion		
	Nominal (US$m)	Per cent (%)	from existing intra-African suppliers (US$m)	from world suppliers (US$m)	Total change in imports (US$m)
Animal and animal products	152	4.9	−23	−81	71
Vegetable products	281	5.4	−34	−184	97
Foodstuffs	411	6.0	−32	−228	183
Mineral products	764	3.2	−51	−552	212
Chemicals and allied industries	432	5.6	−46	−271	161
Plastics/rubbers	287	10.3	−12	−200	88
Raw hides, skins, leather and furs	9	7.8	−0.2	−6	3
Wood and wood products	155	6.4	−13	−94	62
Textiles and apparel	286	10.7	−14	−186	99
Footwear/headgear	43	9.3	−2.1	−27	16
Stone/glass	222	3.0	−88	−94	127
Metals	535	5.0	−25	−354	181
Machinery/electrical	759	12.3	−21	−572	187
Transportation	427	6.7	−118	−255	173
Miscellaneous	194	15.5	−5	−137	57
TOTAL	**4957**	**5.7**	**−485**	**−3241**	**1716**

Notes: The model and data used to generate these results are outlined in Appendix A. The results show the impact of full liberalisation across all products and countries, rather than make assumptions about the products that some countries may exclude from liberalisation. Elasticity sensitivity analysis is shown in Appendix B.

general equilibrium models of the IMF (2019), the World Bank (2020; 2022) and the ECA (2021).[4]

Not all increases in trade brought about by the AfCFTA are 'new'. In many instances, the intra-African trade boosted by the AfCFTA arises from trade diverted away from other suppliers – what is known as 'trade diversion'. This includes both suppliers within the continent (such as intra-REC trade that would now face competition from other African suppliers from outside these RECs) and 'world suppliers' from outside the continent.

Table 6.4 shows the estimated degree of trade diversion. This is the amount of trade that switches from a previous importing partner to a new partner as a result of the change in tariffs making new intra-African suppliers more competitive. In this context, trade diversion is split to show the trade that has been diverted away from intra-African and world suppliers, by HS section. For instance, some maize imports from Uganda to Kenya might be replaced by new imports of maize from South Africa as a result of the AfCFTA, which would be counted as trade diversion from existing intra-African suppliers. Other maize imports into Kenya from India might also be replaced by South Africa, which would be considered to have been diverted from world suppliers. Though the AfCFTA is expected to result in trade diversion between African suppliers, this is small.

About two-thirds of the increase is expected to come from trade diverted from outside the continent and the remainder (22 per cent) from trade creation. The relatively small share of trade diversion between African suppliers owes to the relatively small share of African suppliers in current import flows. In other words, there is little pre-existing intra-African trade to be diverted away from.

Consistent with modelling efforts by other authors (ECA 2021; IMF 2019; World Bank 2020; World Bank 2022), this model forecasts that the AfCFTA will stimulate the largest increases in intra-African trade in manufacturing. This importantly helps the AfCFTA to contribute to Africa's structural transformation and industrialisation, an explicit objective of the Agreement Establishing the African Continental Free Trade Area.[5] The forecast impact on the agricultural sector is nevertheless still important, with intra-African trade in foodstuffs, vegetable products and animal and animal products increasing by 6 per cent, 5.4 per cent and 4.9 per cent, respectively.

Impact of the African Continental Free Trade Area on the agriculture sector by value chain segment

A detailed breakdown of the results for each segment of the agriculture value chain is shown in Table 6.5. In absolute terms, the AfCFTA boosts intra-African trade most in the downstream part of the value chain concerned with foods. The gains to trade in food are larger, in absolute terms, than all other parts of the value chain combined. The reason the gains are so much higher in this part of the value chain is that it is the part of the value chain that currently faces the high tariffs, where the value of imports is largest, and where African producers already have some capacity and existing trade flows to scale up.

Intra-African trade gains are smallest, in absolute terms, for trade in inputs and capital. Tariffs are already relatively low in these parts of the value chain so the benefit of tariff liberalisation under the AfCFTA will be less noticeable. Nevertheless, in percentage terms, the impact on agricultural capital is quite large, at 8.2 per cent.

Table 6.5: Impact of the AfCFTA, by segment of the agriculture value chain

| | Increase in intra-African trade | | Trade diversion | | |
	Nominal (US$m)	Per cent (%)	from existing intra-African suppliers (US$m)	from world suppliers (US$m)	Total change in imports (US$m)
All foods	715	5.3	−84	−402	313
Agricultural raw materials	160	7.5	−8	−111	50
Agricultural capital	81	8.2	−3	−62	19
Agricultural inputs	58	2.8	−4	−34	24
TOTAL	1015	5.4	−99	−609	406

Notes: The model and data used to generate these results are outlined in Appendix A. The results show the impact of full liberalisation across all products and countries, rather than make assumptions about the products that some countries may exclude from liberalisation.

Impact by main agricultural products

Table 6.6 further disaggregates results for the agriculture value chain by main products. This helps to identify the key products driving the impact of the AfCFTA.

The most substantial potential for intra-African trade gains is found in the foods part of the value chain. Within this part of the value chain, there is most potential for the AfCFTA to boost intra-African trade in fish and seafood, sugar, fruit, tobacco, preparations of cereals, vegetables, vegetable oils, beverages and dairy. There are also relatively sizeable opportunities for exports of coffee. Among the more value-added products there are important opportunities for exporters of agricultural machinery and fertilisers.

What is also of note is the relatively small impact the AfCFTA is forecast to have on trade in staple/food security crops, including wheat, maize, millet, sorghum and soya beans. This is because these are products for which average tariffs are already low, meaning that tariff reductions resulting from the AfCFTA can only have a minimal effect. Only other interventions, such as reducing non-tariff barriers, could have the potential to have a transformative impact on intra-African trade in such goods.

Table 6.6: Impact of the AfCFTA on the agriculture sector, by main products

| | AfCFTA impact | | | | |
| | Increase in intra-African trade | | Trade diversion | | |
	Nominal (US$m)	Per cent (%)	from existing intra-African suppliers (US$m)	from world suppliers (US$m)	Total change in imports (US$m)
Vegetables	31	4.1	−3	−18	12
Wheat	15	3.6	−8	−8	8
Beef	6	4.2	−0.4	−4	3
Dairy	25	4.7	−1	−15	10
Fish and seafood	147	7.8	−23	−62	85
Fruit	54	7.0	−9	−26	28
Maize	16	3.3	−6	−7	9
Millet	0	1.0	−0.1	−0.1	0
Palm oil	17	2.6	−2	−12	5
Preparations of cereals	48	7.0	−3	−31	17
Sorghum	1	1.8	−0.1	−1	0
Soya beans	0.05	0.1	−0.2	−0.02	0
Vegetable oils	38	8.3	−1	−29	9
Beverages	28	3.2	−3	−12	16
Sugar	107	7.6	−8	−72	35
Citrus fruit	3	5.1	−1	−1	3
Tobacco	48	4.6	−4	−18	30
Nuts	4	10.2	0	−2	2
Poultry	16	10.1	−1	−12	4
Rice	3	0.9	−1	−3	1
Other food	108	4.0	−10	−70	37
All foods	**715**	**5.3**	**−84**	**−402**	**313**

(Continued)

Table 6.6: (Continued)

	AfCFTA impact				
	Increase in intra-African trade		Trade diversion		
	Nominal (US$m)	Per cent (%)	from existing intra-African suppliers (US$m)	from world suppliers (US$m)	Total change in imports (US$m)
Coffee	73	25.9	−1	−58	16
Tea	5	1.3	−1	−3	2
Cocoa	15	7.8	−1	−8	7
Wood	7	1.8	−2	−3	5
Flowers	1	10.2	0	−1	0
Fibres	28	16.4	0	−19	9
Cotton	1	0.5	−1	−1	0
Other agricultural raw materials	29	5.9	−1	−18	11
All agricultural commodities	**160**	**7.5**	**−8**	**−111**	**50**
Machinery	75	10.0	−3	−57	17
Tools	0.2	1.7	−0.02	−0.1	0
Tractors	6	2.7	−0.2	−5	1
All agricultural capital	**81**	**8.2**	**−3**	**−62**	**19**
Fertilisers	42	2.3	−3	−22	20
Insecticides	16	5.6	−0.2	−12	4
All agricultural inputs	**58**	**2.8**	**−4**	**−34**	**24**
TOTAL	**1015**	**5.4**	**−99**	**−609**	**406**

Notes: The model and data used to generate these results are outlined in Appendix A. The results show the impact of full liberalisation across all products and countries, rather than make assumptions about the products that some countries may exclude from liberalisation. Elasticity sensitivity analysis is shown in Appendix B. Table is shaded, with darker cells showing larger values.

6.4 Country-level impacts

This section breaks down the results at the country level, showing the potential impact of the AfCFTA on intra-African imports by the segments of the agriculture value chain. These results are driven by the prevailing structure of tariffs and trade. The greatest impact is seen in countries (and value chain segments) where both existing intra-African trade and tariffs are large. This makes sense; it is exactly *those* tariffs on *that* trade that the AfCFTA will liberalise.

East Africa

In East Africa, the impact of the AfCFTA on the agriculture sector is forecast to be largest in absolute terms in Ethiopia. Ethiopia does not fully implement any of the REC FTAs (it reportedly applies just a 10 per cent reduction to tariffs on imports from Common Market for Eastern and Southern Africa (COMESA) members), and so has much higher average tariffs on intra-African trade than most other countries in the region. It also has, in general, higher tariffs than many other countries and is a relatively large economy by regional standards.

Much of the relatively large forecast increase in imports into Kenya are products from South Africa, including fruits, sugar and agricultural machinery,

Figure 6.2: Forecast increase in intra-African imports as a result of the AfCFTA in Eastern Africa, by country and value chain segment (US$ million)

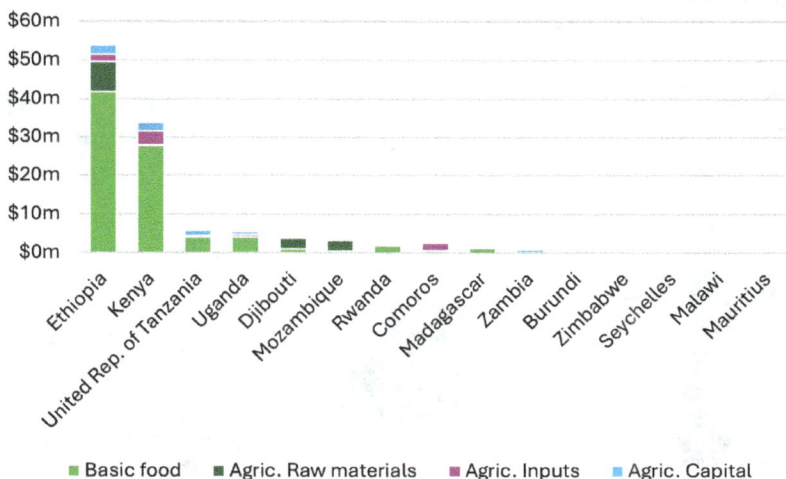

Notes: The model and data used to generate these results are outlined in Appendix A. The results show the impact of full liberalisation across all products and countries, rather than make assumptions about the products that some countries may exclude from liberalisation. We use the regional classification provided by United Nations Department of Economic and Social Affairs (UN DESA n.d.).

but also maize from Egypt. Kenya is a gateway to the region that already has trade flows existing with other parts of the continent. For Tanzania, the main increases in imports are sugar, vegetable oils, preparations of cereals, machinery, and insecticides from Egypt.

The impact of the AfCFTA is forecast to be marginal on agriculture imports for the other countries of East Africa, as shown in Figure 6.2 (which comprises all countries based on the UN definition of East Africa). This is because most of their intra-African trade already occurs through pre-existing REC FTAs (especially the EAC and COMESA) or because, in the case of the Seychelles and Mauritius, they already have very low most-favoured nation (MFN) tariffs on these products.

Central Africa

In Central Africa, the AfCFTA is forecast to have the most potential for increasing agriculture imports into Democratic Republic of the Congo (DRC) and Cameroon, two of the larger markets in the region that both trade with their neighbours outside the Economic Community of Central African States free trade area, as shown in Figure 6.3. A large share of this increase would be imports from South Africa because, while the DRC is a member of the Southern African Development Community (SADC), it does not implement the SADC FTA, which would otherwise cover this trade.

Figure 6.3: Forecast increase in intra-African imports as a result of the AfCFTA in Central Africa, by country and value chain segment (US$ million)

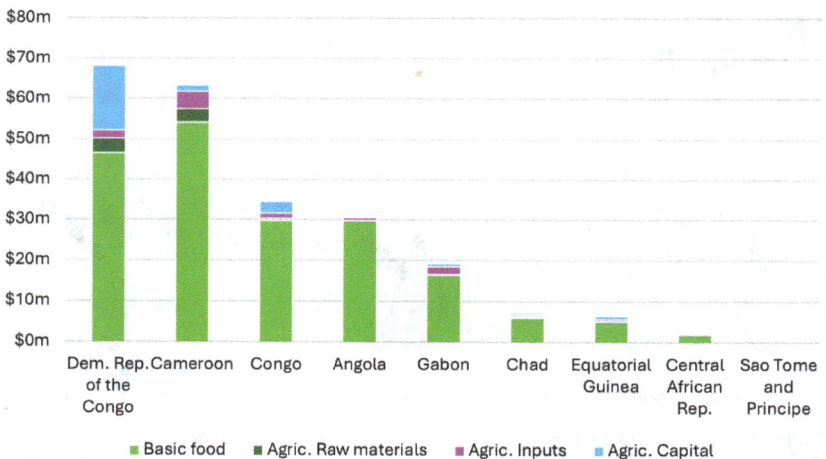

Notes: The model and data used to generate these results are outlined in Appendix A. The results show the impact of full liberalisation across all products and countries, rather than make assumptions about the products that some countries may exclude from liberalisation.

Much of the potential new agriculture imports to Central Africa is fish and seafood, especially from West Africa but also Southern and North Africa. Other important imports are preparations of cereals, vegetables, fruits and sugar, also from Southern and North Africa.

North Africa

In North Africa, most of the potential for increased intra-African agriculture trade is in Algeria (see Figure 6.4). The AfCFTA would see a reduction in tariffs applied by Algeria on products from several of its neighbouring North African countries, but also other countries around the continent, from which it imports coffee, fruits, fish and seafood and tobacco.

For Morocco, Sudan and Tunisia, the increases in imports come from around the continent and involve coffee, fish and seafood, agricultural machinery, fruits, beef, sugar and coffee.

Increased imports into Egypt might seem surprisingly low, given the size of the Egyptian economy. This is because many agriculture imports into Egypt from other African countries are already duty-free, owing to its participation in both the COMESA agreement or the Agadir and Pan-Arab FTA arrangements with its North African neighbours. As such, the AfCFTA does little to boost agriculture imports into Egypt. Libya has very low tariffs to begin with, and so there is little scope for improvements in market access offered through the AfCFTA. As a result, the AfCFTA does little to boost intra-African trade to Libya.

Figure 6.4: Forecast increase in intra-African imports as a result of the AfCFTA in North Africa, by country and value chain segment (US$ million)

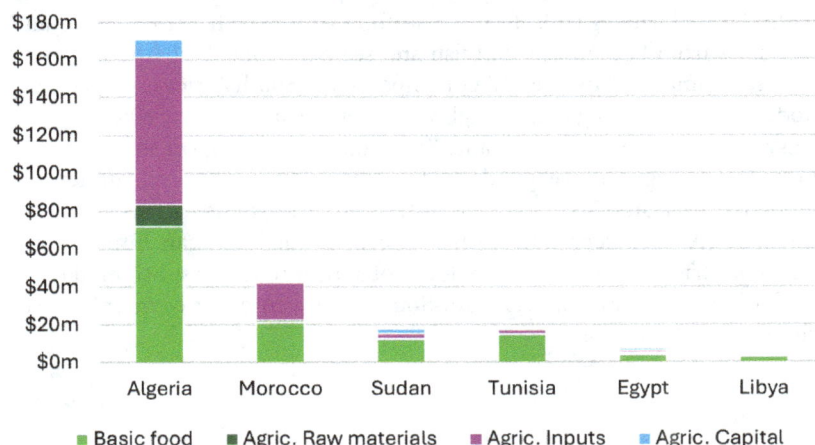

Notes: The model and data used to generate these results are outlined in Appendix A. The results show the impact of full liberalisation across all products and countries, rather than make assumptions about the products that some countries may exclude from liberalisation.

Figure 6.5: Forecast increase in intra-African imports as a result of the AfCFTA in Southern Africa, by country and value chain segment (US$ million)

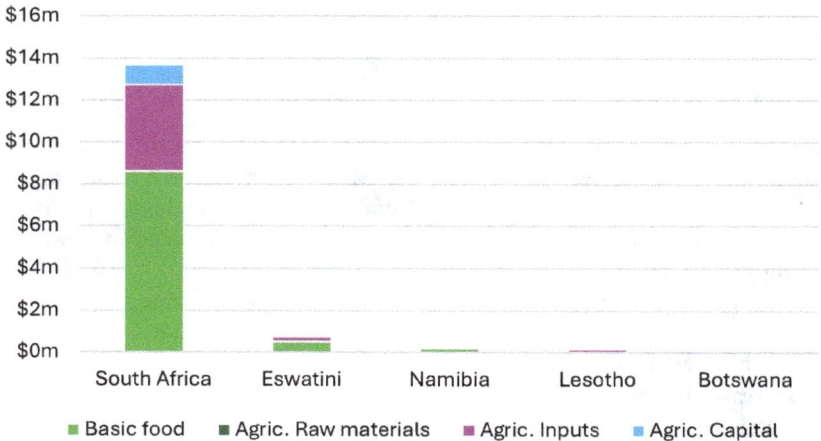

Notes: The model and data used to generate these results are outlined in Appendix A. The results show the impact of full liberalisation across all products and countries, rather than make assumptions about the products that some countries may exclude from liberalisation.

Southern Africa

South Africa is by far the largest economy in the Southern Africa region and unsurprisingly accounts for most of the potential for agriculture imports under the AfCFTA (Figure 6.5). Important potential for new imports into South Africa includes fruit, fish and seafood from North Africa, coffee and vegetables from East Africa, and fish and seafood from West Africa. Nevertheless, the impact of the AfCFTA on imports into South Africa is surprisingly modest, possibly owing to how highly competitive its domestic economy is.

Eswatini, Namibia, Lesotho and Botswana already import most of their intra-African agriculture goods from South Africa duty-free under the Southern African Customs Union trading arrangements. As a result, the AfCFTA does very little to increase their intra-African imports. For these countries, the AfCFTA also does not appear to create substantial trade diversion away from imports from South Africa to other economies elsewhere in the continent.

West Africa

Increases in imports to West Africa driven by liberalised trade with other parts of the continent will be significant (Figure 6.6). Although Nigeria is by far the largest economy in West Africa, increases in intra-African imports

Figure 6.6: Forecast increase in imports as a result of the AfCFTA in West Africa, by country and value chain segment (US$ million)

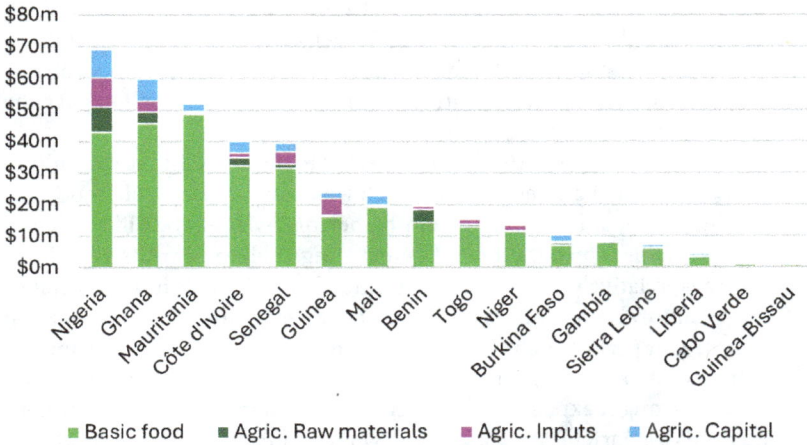

Legend: ■ Basic food   ■ Agric. Raw materials   ■ Agric. Inputs   ■ Agric. Capital

Notes: The model and data used to generate these results are outlined in Appendix A. The results show the impact of full liberalisation across all products and countries, rather than make assumptions about the products that some countries may exclude from liberalisation.

are only forecast to be marginally larger for Nigeria than for other West African countries, such as Ghana, Mauritania, Côte d'Ivoire and Senegal. This stems from the pre-existing intra-African trade flows, which are relatively limited for Nigeria. Most Nigerian food imports are sourced from outside the continent, for instance.

Important new intra-Africa imports into West Africa include sugar, fruits, dairy, vegetables and agricultural machinery from South Africa, and dairy, fish and seafood and fertiliser from North Africa.

Summary

A detailed partial equilibrium model was used to simulate the impact of the AfCFTA. This allowed its effects to be forecast at a highly detailed level of disaggregation to show likely implications for different segments of the agriculture value chain and for specific products.

The impact of the AfCFTA on intra-African trade is relatively modest. That is because much of that trade is already liberalised through pre-existing subregional trade agreements across the continent, such as those of the EAC, COMESA, SADC, ECOWAS, the Pan-Arab FTA and the Agadir Agreement. It is through these subregional arrangements that most of Africa's *current* intra-African trade in the agriculture sector flows.

What the AfCFTA is really doing is liberalising the (currently) smaller shares of intra-African trade that flow *between* regions, such as from Southern Africa to West Africa, or North Africa to Eastern Africa. Tariffs on these goods are high, averaging 18 per cent for vegetable products and 19 per cent for foodstuffs, for instance. It is the liberalisation of this trade that drives the modelled estimates and for which we will expect the AfCFTA to have most impact.

Where the impact of the AfCFTA is expected to be largest in the agriculture sector, in the immediate term, is in the downstream consumable food part of the value chain, and especially with higher-unit-value foods like fish and seafood, vegetables, preparations of cereals, vegetable oils, fruits and dairy. There are also relatively sizeable opportunities for exports of sugar and coffee, within agricultural commodities. Though the opportunities for trade creation in the upstream part of the value chain are smaller in total, there are important opportunities for exporters of agricultural machinery, fertilisers and pesticides. We might expect South Africa to begin supplying more of the continent's needs for agricultural machinery, while more fertilisers and pesticides could be expected from North Africa.

The AfCFTA is likely to have less of an impact on trade in staple/food security crops, including wheat, maize, rice, millet, sorghum and soya beans. These products already have, on average, low tariffs or are traded through informal cross-border trade, as well as by suppliers outside the continent. As a result, the AfCFTA is expected in the short term to have little direct impact on food security through an accessibility channel unless it can go beyond merely reducing tariffs. To improve food security, the AfCFTA will need to do more to address non-tariff barriers, attract investments, and facilitate a broader coordination of relevant policies.

It is also worth raising an inherent limitation of almost all ex ante trade models, which is that they must (necessarily) be fed with data on current trade flows. They are able to scale up, and down, those trade flows to show where demand is created and substituted between import suppliers. However, they are unable to simply create new trade flows where they did not previously exist. This inherent feature of such modelling might be compared to driving looking only in the rear-view mirror, failing to see a possible turning in the road ahead. Identifying, and seizing, such wholly new opportunities would be at the heart of a more impactful AfCFTA on the agriculture sector and will require bold vision by African leaders.

The main conclusion is exactly that. If the AfCFTA is to have a transformative impact on Africa's agriculture sector it must entail much more than just tariff liberalisation (though tariff liberalisation is a starting point that would certainly help). The AfCFTA will need to stimulate the creation of wholly new patterns of trade through enticing investments, coordinating policies and addressing non-tariff barriers, which are often more burdensome than merely tariffs for agricultural trade. Part of the solution can also entail leveraging informal cross-border trade, which exists in substantial quantities (Gaarder, Luke and Sommer 2021) but is, by its definition, unrecorded and does not flow between countries through typical formal trade routes. Chapter 5 discussed the magnitude of such trade.

Notes

[1] Article 19, paragraph 2 reads: 'State Parties that are members of other regional economic communities, regional trading arrangements and custom unions, which have attained among themselves higher levels of regional integration than under this Agreement, shall maintain such higher levels among themselves.'

[2] Note that when aggregated, as in Table 6.1, import-weighted tariffs may underestimate the restrictiveness of the tariffs when comparing different products (since the variance of the tariffs and the import demand elasticities can be different within each grouping). Intuitively, this owes to businesses importing less of products that are tariffed highly. That does not, however, affect the underlying modelling (which is undertaken at the more disaggregated HS-6 level, where tariffs are not aggregated to this extent).

[3] However, other measures that are foreseen in the AfCFTA (tackling non-tariff barriers and improving preference utilisation) could have a significant effect, even where tariffs are low (De Melo, Sorgho and Wagner 2023; United Nations Conference on Trade and Development 2019).

[4] In fact, heroic assumptions about non-tariff barriers, trade facilitation and other measures, account for almost 97.5 per cent of the estimated impact of the AfCFTA in the models of the World Bank (2022) and the IMF (2019).

[5] See Article 3 (e) and (g) of the Agreement Establishing the African Continental Free Trade Area.

References

Andriamananjara, Soamiely; Brenton, Paul; von Uexküll, Jan Erik; Walkenhorst, Peter (2009) 'Assessing the Economic Impacts of an Economic Partnership Agreement on Nigeria', World Bank. https://perma.cc/M98Z-E9W5

Agreement Establishing the African Continental Free Trade Area, 2018.

Bayale Nimonka; Ibrahim, Muazu; and Atta-Mensah, Joseph (2022) 'Potential Trade, Welfare and Revenue Implications of the African

Continental Free Trade Area (AfCFTA) for Ghana: An Application of Partial Equilibrium Model', *Journal of Public Affairs*, vol. 22, no. 1. https://doi.org/10.1002/pa.2385

Brenton, Paul; Hoppe, M.; and von Uexkull, Jan Erik (2007) 'Evaluating the Revenue Effects of Trade Policy Options for COMESA Countries: The Impacts of a Customs Union and an EPA with the European Union', The World Bank, Technical Report.

UN DESA (n.d.) 'Classification and Definition of Regions', United Nations Department of Economic and Social Affairs. https://esa.un.org/MigFlows/Definition%20of%20regions.pdf

CEPII (n.d.) 'Description of BACI', CEPII. https://perma.cc/EH9C-YH5X

De Melo, Jaime; Sorgho, Zakaria; and Wagner, Laurent (2023) 'DP18651 Reducing Wait Times at Customs: How Implementing the Trade Facilitation Agreement (TFA) Can Expand Trade among AfCFTA Countries', CEPR Discussion Papers, 18651. https://perma.cc/CT4M-TDKH

Dimaranan, Betina V.; McDougall, Robert A.; and Hertel, Thomas W. (2006) 'Behavioral Parameters', in *Global Trade, Assistance, and Production: The GTAP 6 Data Base*, Center for Global Trade Analysis, Purdue University. https://perma.cc/NMJ8-RZYT

ECA (2021) 'New Assessment of the Economic Impacts of the Agreement Establishing the African Continental Free Trade Area on Africa', United Nations Economic Commission for Africa. https://repository.uneca.org/bitstream/handle/10855/46750/b11999160.pdf?sequence=1&isAllowed=y

ECA, AU and AfDB (2017) *Assessing Regional Integration in Africa VIII: Bringing the Continental Free Trade Area About.* Addis Ababa, Ethiopia: United Nations Economic Commission for Africa, African Union and African Development Bank. https://www.un-ilibrary.org/content/books/9789213615591.

Fouda Ekobena, Simon Yannick; Coulibaly, Adama Ekberg; Keita, Mama; and Antonio, Pedro (2021) 'Potentials of the African Continental Free Trade Area: A Combined Partial and General Equilibrium Modeling Assessment for Central Africa', *African Development Review*. https://doi.org/10.1111/1467-8268.12594

Gaarder, Erwin; Luke, David; and Sommer, Lily (2021) 'Towards an Estimate of Informal Cross-Border Trade in Africa', United Nations Economic Commission for Africa. https://repository.uneca.org/handle/10855/46374#:~:text=We%20estimate%20that%20the%20value,formal%20trade%20between%20neighbouring%20countries

IMF (2019) 'The African Continental Free Trade Agreement: Welfare Gains Estimates from a General Equilibrium Model', IMF. https://perma.cc/S9GH-NFSY

ITC (2015) 'The Invisible Barriers to Trade. How Businesses Experience Non-tariff Measures', Technical Paper, Doc. No. MAR-15-326.E, Geneva: INTRACEN. https://perma.cc/6AJ7-SV3J

Lunenborg, Peter; and Roberts, Thomas (2021) 'ECOWAS and AfCFTA: Potential Short-Run Impact of a Draft ECOWAS Tariff Offer', *Journal of African Trade*, vol. 8, no. 2. https://doi.org/10.2991/jat.k.211011.001

MacLeod, Jamie; and Von Uexkull, Jan Erik (2016) 'Assessing the Economic Impact of the ECOWAS CET and Economic Partnership Agreement on Ghana', World Bank. https://perma.cc/2ZS2-KVAL

Mulugeta, Tages (2020) 'The Revenue Implications of African Continental Free Trade Area (AFCFTA): In Case of Ethiopia' (Doctoral dissertation, St. Mary's University).' http://197.156.93.91/handle/123456789/6676

Ossadzifo Wonyra, Kwami; and Bayale, Nimonka (2022) 'Assessing the Potential Effects of The AfCFTA on Togolese Economy: An Application of Partial Equilibrium Model', *Journal of Public Affairs*, vol. 22, no. 1. https://doi.org/10.1002/pa.2377

Oyelami, Lukman O. (2021) 'Revenue, Welfare and Trade Effects of African Continental Free Trade Agreement (AFCFTA) on Nigerian Economy', *Journal of Public Affairs* [Preprint]. https://doi.org/10.1002/pa.2645

Seti, Thembalethu M.; and Daw, Olebogeng D. (2022) 'The Implications of the African Continental Free Trade Area on South African Agricultural trade: an Application of the Partial Equilibrium Mode', *South African Journal of Economic and Management Sciences*, vol. 25, no. 1. https://doi.org/10.4102/sajems.v25i1.4302

United Nations Conference on Trade and Development (2019) 'Economic Development in Africa Report 2019 Made in Africa Rules of Origin for Enhanced Intra-African Trade', Geneva: United Nations Conference on Trade and Development. https://perma.cc/LJK2-6UR9

World Bank (2020) 'The African Continental Free Trade Area: Economic and Distributional Effects', World Bank. https://hdl.handle.net/10986/34139

World Bank; Echandi, R. Roberto; Maliszewska, Maryla; and Steenbergen, Victor (2022) 'Making the Most of the African Continental Free Trade Area : Leveraging Trade and Foreign Direct Investment to Boost Growth and Reduce Poverty', World Bank. https://perma.cc/KLL9-VETY

Appendix A. Model, data and reform scenario

Reform scenario

The reform scenario simulates the AfCFTA with a focus on the agriculture sector. In so doing, it intends to show the potential of tariff liberalisation under the AfCFTA, rather than other aspects such as decisions over exclusion lists, trade facilitation assumptions, reductions in non-tariff barriers, or efforts in the areas of trade in services, investment, competition policy, intellectual property rights or other areas.[1] These supplementary aspects require stronger assumptions and researchers use a different approach to model them.

The reform scenario reflects full implementation of the AfCFTA once the complete course of any incremental tariff reductions has been applied. It applies tariff liberalisation to all goods rather than make assumptions over sensitive product and exclusion lists.

What trade models like that used here can do is to scale up, or down, existing trade flows in proportion to changes in other variables such as tariffs. They cannot create wholly new trade flows from nothing. As such, modelling exercises such as this one may fail to identify where brand new trade flows may emerge between trading partners that did not previously trade certain products. That is more likely to happen in instances where a trade agreement results in a very large change in some tariffs.

Structure of the dataset

Trade flows data is taken from the BACI dataset of reconciled trade flows prepared by CEPII, which is in turn based on data reported by countries to the United Nations Statistical Division Comtrade dataset.

Both exporting and importing countries report data for Comtrade. The CEPII-BACI dataset reconciles these two mirror sources of reported trade data into a single dataset. This is done through an approach that reflects the reliability of different reports of the same trade flows while stripping out insurance and freight costs to express all trade data in terms of their free-on-board price. Doing so uses all available information to maximise data coverage in instances where reporting may be incomplete or of varying qualities of reliability. This is particularly valuable in trade, such as intra-African trade, that comprises flows between less-developed countries, many of which have less well-resourced data collection systems in place. It also makes our work easier and results more intuitive; what Ghana exports to Kenya becomes exactly which Kenya imports from Ghana.

A three-year average of trade flows from 2017 to 2019 is used. These years are the most recent consecutive three-year period that can be considered to represent 'normal' trade flows unaffected by the economic volatility of the Covid-19 pandemic. These were also the three years at the time of writing with the highest number of observations (distinct combinations of importer–exporter–product with at least one non-zero trade flow) in the BACI dataset (CEPII n.d.), indicating superior reported data coverage.

HS revision 2002 was used. Why not the more recent HS revisions 2007, 2017 or 2022? A number of African countries do not yet report trade flows data in more recent HS revisions, meaning that they are excluded from data that includes only more recent formats (CEPII n.d.). Using an older revision allows the maximum amount of reported trade data to be used in the analysis.[2] HS revisions are backward-compatible, meaning that data captured in more recent versions can be transposed into older versions (but not vice versa).

The CEPII-BACI dataset that was chosen for this study results in comprehensive coverage of countries and products and relatively reliable coverage for countries with less well-resourced reporting systems. It allows analysis at the subheading level of the HS, which in turn allows its reconstitution into appropriate levels of aggregation for the presentation of our results, including at each segment of the agriculture value chain.

Tariff data is drawn from two sources. Where available, data was taken from countries' submissions to the WTO integrated database of applied tariffs for all WTO members as well as some countries that have submitted tariff information to the WTO but are not WTO members, for instance during ongoing accession negotiations.

The most recent year of submitted tariff data was used for each country. Typically, this was for the year 2020 or 2021, allowing a highly up-to-date analysis of tariff information, although, where unavailable, older tariff schedules were used for a few countries. Such data was available in the HS 2017 nomenclature for 43 countries and in earlier nomenclatures for a further four countries. UN Trade Statistics correspondence tables were used to convert all tariff schedules into the 2002 revision in alignment with the trade flows data used.

Not all members of the AfCFTA are members of the WTO or have otherwise submitted tariff schedules to be reported in the WTO integrated database. Tariff data for a further four countries was taken from ITC's MAcMap tariff database.[3]

No publicly available tariff data was available for four AfCFTA participating countries (Eritrea, Sahrawi Republic, Somalia and South Sudan). The impact of the AfCFTA on imports into these countries could therefore not be calculated. However, exports from these countries into other AfCFTA member countries is captured and included in the analysis through mirror reporting.

Model specification

In order to calculate the percentage change in the price of good k from exporter i due to a change in tariff t, the model uses the following formula:

$$\frac{\Delta p_i}{p_i^{old}} = \frac{\left[\frac{p_i^{new}}{P_{wld}}\right] - \left[\frac{p_i^{old}}{P_{wld}}\right]}{\frac{p_i^{old}}{P_{wld}}} = \frac{\left(1+t_i^{new}\right) - \left(1+t_i^{old}\right)}{\left(1+t_i^{old}\right)} = \frac{t_i^{new} - t_i^{old}}{\left(1+t_i^{old}\right)}$$

where superscripts 'new' and 'old' denote the prices and tariffs before and after the policy reform.

The import response is calculated in two consecutive steps. The first step is the substitution between different exporters due to changes in their relative tariff rates. A given expenditure for imports of good k is reallocated across different exporters following the change in relative prices as follows:

$$M_i^{ES} = \left[\frac{\Delta p_i}{p_i^{old}} \gamma_i^{ES} + 1\right] M_i^{old} \frac{\sum_{i=1,...,n} M_i^{old}}{\sum_{i=1,...,n}\left(\left[\frac{\Delta p_i}{p_i^{old}} \gamma_i^{ES}\right] M_i^{old}\right)}$$

where M_i^{ES} stands for the imported quantity from i after exporter substitution, M_i^{old} is the imported quantity from i before reform, and γ_i^{ES} is the exporter substitution elasticity for imports from country i.

The second step is the demand effect. It depends on the price change for the total basket of imports $\overline{P}$, as a result of the price change on imports from country i, which is given by:

$$\frac{\Delta \overline{P}}{\overline{P}^{old}} = \sum_{i,...,n}\left[\frac{M_i^{old}}{\sum_{i,...,n} M_i^{old}} \frac{\Delta p_i}{p_i^{old}}\right]$$

which, through the elasticity of demand μ^D, leads to a change in the total demand for imports from all sources M^{ED}.

$$M^{ED} = \left[\frac{\Delta \overline{P}}{\overline{P}^{old}} \mu^D + 1\right] M^{old}$$

resulting in the new import quantity M_i^{new} from country i as follows:

$$M_i^{new} = M_i^{ES} + \left[M^{ED} - M^{old}\right]\left[\frac{M_i^{old}}{\sum_{i,...,n} M_i^{old}}\right]$$

Structural parameters

The values of exporter substitution and demand elasticities are subject to some uncertainty. Three versions of the model were therefore prepared using different values for these parameters. In the first 'low elasticity' model, lower end estimates of exporter and demand elasticities are used. In it, importers are less sensitive in their sourcing decisions to tariff-price changes. This results in a much smaller estimated impact of the AfCFTA. A second 'high elasticity' model was developed for comparison. Finally, a third model relying on the elasticities used in the standard Global Trade Analysis Project (GTAP) model was developed. In this third model, exporter substitution elasticities vary across sectors and demand elasticities change depending on the country that is doing the importing, which is more realistic. The GTAP elasticity parameters are closer to, and in fact on average exceed those of, the 'high elasticity model'. This is because they consider a longer time horizon in which consumer decisions have had better chance to adjust to changing prices.

The 'low elasticity' and 'high elasticity' models benefit in that their results are determined entirely by differences in the structure and shape of tariffs and trade flows, rather than assumptions over relative differences between products and countries' elasticity parameters (since these are uncertain, using a model in which they drive the results in different countries could lead to erroneous conclusions being drawn). Their results might be argued to be more transparent and are used in different ways by a number of authors (Brenton, Hoppe and von Uexkull 2007; ECA, AU and AfDB, 2017; Lunenborg and Roberts 2021; MacLeod and von Uexkull 2016; Andriamananjara et al. 2009).[4] However, to improve relative comparability with most of the existing literature, the results in this paper (unless otherwise specified) rely on the third model, which uses the GTAP elasticities. These have the advantage of more realistically varying by product and importer country though at the cost of making the model somewhat more complex and less intuitive. Comparative results for the 'low elasticity' and 'high elasticity' models are included in Appendix B and details of all elasticity parameters included in Appendix C.

In the partial equilibrium model, a preferential liberalisation of a given tariff affects not only the overall price level of the good but also the relative prices of the different varieties. Through the import demand elasticity and the substitution elasticity, this will lead to changes in the aggregate level of spending on that good, as well as changes in the composition of the sourcing of that good. Both channels affect bilateral trade flows. The model estimates the potential impact of a given tariff reform scenario on both source specific and total imports, at the HS 6-digit level. This level of disaggregation reduces the risk of biases in calculating and operating with *average* tariff rates across groups of products and allows the results to be reconstituted into intuitive product categories for the value chain and for the decisions that negotiators are making.

Appendix B. Elasticity sensitivity analysis: comparative low and high elasticity models, by HS section

Table 6.7: Low elasticity parameters: Impact of the AfCFTA, by HS section

	AfCFTA impact				
	Increase in intra-African trade		Trade diversion		
	Nominal (US$m)	Per cent (%)	from existing intra-African suppliers (US$m)	from world suppliers (US$m)	Total change in imports (US$m)
Animal and animal products	63	2	−8	−25	38
Vegetable products	117	2	−12	−47	69
Foodstuffs	190	3	−13	−80	111
Mineral products	160	1	−10	−73	87
Chemicals and allied industries	120	2	−10	−57	63
Plastics/rubbers	74	3	−2	−41	33
Raw hides, skins, leather and furs	2	2	−0.1	−1	1
Wood and wood products	46	2	−3	−22	24
Textiles and apparel	73	3	−2	−35	38
Footwear/headgear	11	2	−0.4	−5	6
Stone/glass	69	1	−17	−22	48
Metals	136	1	−5	−67	69
Machinery/electrical	158	3	−3	−90	68
Transportation	122	2	−20	−54	68
Miscellaneous	45	4	−1	−24	20
TOTAL	**1385**	**2**	**−106**	**−642**	**743**

Notes: Low elasticities: substitution elasticity = 1.5, demand elasticity = 0.5, High elasticities: substitution elasticity = 5, demand elasticity = 1. See Andriamananjara et al. (2009).

Table 6.8: High elasticity parameters: impact of the AfCFTA, by HS section

| | AfCFTA impact | | | | |
| | Increase in intra-African trade | | Trade diversion | | |
	Nominal (US$m)	Per cent (%)	from existing intra-African suppliers (US$m)	from world suppliers (US$m)	Total change in imports (US$m)
Animal and animal products	167	5	−28	−84	83
Vegetable products	315	6	−39	−159	156
Foodstuffs	522	8	−43	−275	247
Mineral products	458	2	−35	−276	182
Chemicals and allied industries	344	5	−35	−204	140
Plastics/rubbers	225	8	−9	−150	75
Raw hides, skins, leather and furs	6	6	−0.3	−4	2
Wood and wood products	132	5	−11	−79	54
Textiles and apparel	211	8	−9	−123	88
Footwear/headgear	30	7	−1.4	−17	13
Stone/glass	181	2	−57	−75	106
Metals	397	4	−18	−243	154
Machinery/electrical	490	8	−12	−336	153
Transportation	348	5	−71	−198	149
Miscellaneous	136	11	−3	−89	47
TOTAL	**3962**	**5**	**−371**	**−2312**	**1650**

Notes: Low elasticities: substitution elasticity = 1.5, demand elasticity = 0.5, High elasticities: substitution elasticity = 5, demand elasticity = 1. See Andriamananjara et al. (2009).

Appendix C. Elasticity parameters

Table 6.9: Substitution elasticity parameters

GTAP Sector	Description	GTAP Substitution elasticity (γ)	Low substitution elasticity (γ)	High substitution elasticity (γ)
pdr	Paddy rice	10.1	1.5	5
wht	Wheat	8.9	1.5	5
gro	Cereal grains n.e.c.	2.6	1.5	5
v_f	Vegetables, fruit, nuts	3.7	1.5	5
osd	Oil seeds	4.9	1.5	5
c_b	Sugar cane, sugar beet	5.4	1.5	5
pfb	Plant-based fibres	5	1.5	5
ocr	Crops n.e.c.	6.5	1.5	5
ctl	Bovine cattle, sheep and goats,	4	1.5	5
oap	Animal products n.e.c.	2.6	1.5	5
rmk	Raw milk	7.3	1.5	5
wol	Wool, silk-worm cocoons	12.9	1.5	5
frs	Forestry	5	1.5	5
fsh	Fishing	2.5	1.5	5
coa	Coal	6.1	1.5	5
oil	Oil	10.4	1.5	5
gas	Gas	34.4	1.5	5
omn	Minerals n.e.c.	1.8	1.5	5
cmt	Bovine meat prods	7.7	1.5	5
omt	Meat products n.e.c.4.40	8.8	1.5	5
vol	Vegetable oils and fats	6.6	1.5	5
mil	Dairy products	7.3	1.5	5
pcr	Processed rice	5.2	1.5	5
sgr	Sugar	5.4	1.5	5
ofd	Food products n.e.c.	4	1.5	5
b_t	Beverages and tobacco products	2.3	1.5	5
tex	Textiles	7.5	1.5	5

GTAP Sector	Description	GTAP Substitution elasticity (γ)	Low substitution elasticity (γ)	High substitution elasticity (γ)
wap	Wearing apparel	7.4	1.5	5
lea	Leather products	8.1	1.5	5
lum	Wood products	6.8	1.5	5
ppp	Paper products, publishing	5.9	1.5	5
p_c	Petroleum, coal products	4.2	1.5	5
crp	Chemical, rubber, plastic products	6.6	1.5	5
nmm	Mineral products n.e.c.	5.8	1.5	5
i_s	Ferrous metals	5.9	1.5	5
nfm	metals n.e.c.	8.4	1.5	5
fmp	Metal products	7.5	1.5	5
mvh	Motor vehicles and parts	5.6	1.5	5
otn	Transport equipment n.e.c.	8.6	1.5	5
ele	Electronic equipment	8.8	1.5	5
ome	Machinery and equipment n.e.c.	8.1	1.5	5
omf	Manufactures n.e.c.	7.5	1.5	5
ely	Electricity	5.6	1.5	5
Average		**7.0**	**1.5**	**5**

Note: GTAP 6 elasticity parameters available from Dimaranan, McDougall and Hertel (2006).

Table 6.10: Demand elasticity parameters

GTAP Country/region	GTAP demand elasticities (μ)							Low demand elasticity (μ)	High elasticity (μ)
	GrainCrops	MeatDairy	OthFoodBev	TextAppar	HouseUtils	Mnfcs	TransComm		
EGY	0.61	0.87	0.79	0.92	0.97	1.07	0.99	0.5	1
MAR	0.56	0.88	0.83	0.94	1.01	1.15	1.03	0.5	1
TUN	0.43	0.78	0.81	0.87	0.96	1.15	0.99	0.5	1
XNF	0.52	0.8	0.8	0.88	0.95	1.12	0.98	0.5	1
NGA	0.52	1.28	0.68	1.29	1.31	1.11	1.32	0.5	1
SEN	0.65	0.94	0.81	0.98	1.05	1.09	1.04	0.5	1
XWF	0.68	1.07	0.82	1.09	1.12	1.12	1.14	0.5	1
XCF	0.66	0.97	0.82	1.01	1.05	1.12	1.07	0.5	1
XAC	0.59	1.28	0.7	1.3	1.31	1.11	1.32	0.5	1
ETH	0.37	1.8	0.44	1.81	1.82	1	1.82	0.5	1
MDG	0.58	1.23	0.71	1.25	1.26	1.07	1.27	0.5	1
MWI	0.52	1.41	0.66	1.42	1.43	1.11	1.44	0.5	1
MUS	0.31	0.72	0.79	0.82	0.92	1.09	0.95	0.5	1
MOZ	0.64	1.2	0.76	1.21	1.23	1.12	1.24	0.5	1
TZA	0.69	1.28	0.84	1.3	1.33	1.22	1.34	0.5	1
UGA	0.65	1.34	0.79	1.36	1.37	1.19	1.38	0.5	1
ZMB	0.68	1.1	0.82	1.12	1.15	1.13	1.16	0.5	1
ZWE	0.57	1.15	0.69	1.17	1.18	1.04	1.19	0.5	1
XEC	0.67	1.08	0.82	1.1	1.14	1.14	1.15	0.5	1
BWA	0.46	0.75	0.77	0.83	0.9	1.07	0.93	0.5	1
ZAF	0.31	0.69	0.77	0.8	0.89	1.06	0.92	0.5	1
XSC	0.58	0.82	0.78	0.87	0.93	1.05	0.95	0.5	1
Average	0.56	1.07	0.76	1.11	1.15	1.11	1.16	0.5	1

Notes (appendices)

[1] Though several states have now submitted tariff schedules under the AfCFTA, not all have, so it would not make sense to apply tariff schedules for only some countries.

[2] In test results using HS17, as much as a fifth of intra-African trade was missing from the data as compared to the results using HS02, for example.

[3] These countries are Ethiopia, Libya, Sao Tome and Principe and Sudan.

[4] Lunenborg and Roberts (2021) use product-specific demand elasticities but common exporter substitution elasticities.

7. Food security in the African Continental Free Trade Area legal framework

Colette Van der Ven

The AfCFTA is expected to play a catalysing role in bringing about more intra-African agriculture and food trade. But, as we saw from the findings of the partial equilibrium modelling exercise in Chapter 6, its overall impact on intra-African food trade is projected to be modest. While the AfCFTA legal instruments contain only minimal references to food security, implementation of provisions on non-tariff barriers (NTBs) in its annexes and protocols can have a stronger impact on achieving food security outcomes across the continent.[1] As we also saw in Chapter 6, attending to NTBs will bring about substantial gains to intra-African food trade, unlike reductions in tariffs since these are already relatively low, thanks to trade liberalisation within the continent's regional economic communities (RECs).

It is to this end that we review the provisions in the AfCFTA on NTBs. The first part of the chapter highlights where the AfCFTA Agreement, protocols and annexes explicitly refer to food security and agriculture and what these provisions entail. Comparisons are made with the WTO Agreement on Agriculture, which is discussed in Chapter 9. The second part of the chapter turns the spotlight on NTB provisions in the AfCFTA legal instruments. Finally, in line with an underlying theme of this book that considers the intersection between trade, food security and climate, the third part of the chapter considers environmental provisions in the AfCFTA legal framework.

7.1 Food security provisions in the African Continental Free Trade Area

Direct references to food security in the African Continental Free Trade Area legal instruments

The AfCFTA legal instruments consist of the Agreement Establishing the African Continental Free Trade Area ('the Agreement') and various protocols,

How to cite this book chapter:

Van der Ven, Colette (2025) 'Food security in the African Continental
   Free Trade Area legal framework', in: Luke, David (ed) *How Africa Eats:
   Trade, Food Security and Climate Risks,* London: LSE Press, pp. 159–186.
   https://doi.org/10.31389/lsepress.hae.g License: CC-BY-NC 4.0

covering trade in goods, services and dispute settlement, as well as competition policy, intellectual property, investment, e-commerce and women and youth (see Box 7.1). The Protocols on Trade in Goods and Trade in Services are accompanied by several annexes, but notably there is no annex dedicated to agriculture or food security. However, the legal instruments that comprise the AfCFTA make various direct – and indirect – references to food security.

The preamble to the Agreement reaffirms the commitment of the member states[2] to the aspirations of Agenda 2063, which includes boosting food security. More substantively, Article 3 (g) of the Agreement specifies that promoting agricultural development and food security is one of the objectives of the AfCFTA (Agreement Establishing the African Continental Free Trade Area, 2018, art 3 (g)). While these references signal that food security is an important objective of the AfCFTA, they do not establish legally binding obligations.

The AfCFTA Protocol on Intellectual Property Rights also contains direct references to food security, without conferring a legal obligation. Article 4, which sets out 'general guiding principles', highlights the 'promotion of the public interest in sectors of vital importance to socio-economic and technological development, including … agriculture, food security and nutrition' (Protocol to the Agreement Establishing the African Continental Free Trade Area on Intellectual Property Rights, 2023, art 4). The section that sets out priority areas of cooperation also provides for 'facilitating the use

Box 7.1: Overview of the architecture of the AfCFTA

The AfCFTA architecture consists of the Agreement and a set of protocols, some accompanied by annexes, adopted in relation to negotiation phases. Phase I (concluded in 2018 and which entered into force in 2019 following ratification by the required number of member states) is made up of the Protocol on Trade in Goods, the Protocol on Trade in Services and the Protocol on Rules and Procedures on the Settlement of Disputes – each of them accompanied by various annexes. At the time of writing, 98 per cent of the negotiations on rules of origin had been completed. Tariff schedules and specific schedules of commitment in services had been completed for almost all member states. Phase II (concluded in 2023 but which has not entered into force, with the required number of ratifications outstanding at the time of writing) consists of Protocols on Competition Policy, Intellectual Property Rights, and Investment. Phase III comprises a Protocol on Digital Trade and a Protocol on Women and Youth (concluded in 2024 but which have also not entered into force, with ratification outstanding at the time of writing). A diagrammatic representation of the AfCFTA's legal architecture is provided in Figure 7.1.

Figure 7.1: The AfCFTA's legal framework and phases

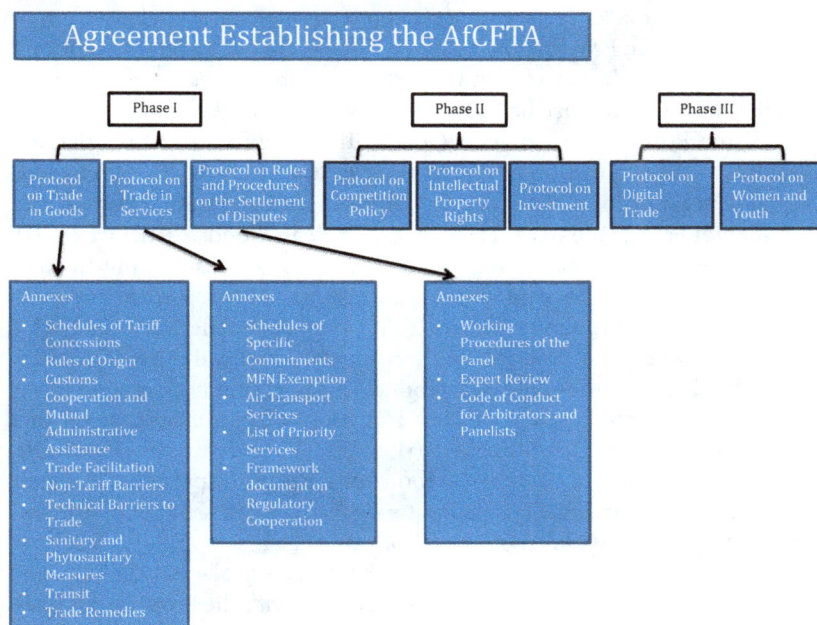

Source: This figure is modified and updated, based on a figure originally published by Tralac in "African Continental Free Trade Area: Questions and Answers." Available at: https://www.tralac.org/documents/resources/faqs/2377-african-continental-free-trade -area-faqs-june-2018update/file.html.

of flexibilities under international instruments for the protection of public health, food security, agriculture and nutrition' (Protocol to the Agreement Establishing the African Continental Free Trade Area on Intellectual Property Rights, 2023, art 23 (f)). Similar to the references to food security in the Agreement, these provisions signal the important link between food security objectives and the AfCFTA but are not enforceable.

Other provisions have more teeth. For example, the Protocol on Trade in Goods contains stipulations that allow member states to take measures to advance food security that would otherwise be inconsistent with the AfCFTA trade liberalisation objectives. For instance, Article 9 allows member states to introduce quantitative restrictions on imports and exports if needed, incorporating by reference Article XI of the General Agreement on Tariffs and Trade (GATT) (see Chapter 8 on the WTO).

However, and again similar to Article XI of GATT, it exempts 'export prohibitions or restrictions temporarily applied to prevent or relieve critical shortages of foodstuffs or other products essential to the exporting party' (GATT 1947, art XI.2(a)). As is discussed in more detail in Chapter 8, while export restrictions can temper domestic price increases, they also risk accelerating price spikes that can have a broader destabilising effect on international markets.

Article 26 of the Protocol on Trade in Goods sets out general exceptions, which allows member states, under certain circumstances, to adopt measures that would otherwise be inconsistent with the trade liberalising objectives of the Protocol on Trade in Goods. This includes measures that are 'essential to the acquisition or distribution of foodstuffs or any other products in general or local short supply' (Agreement Establishing the African Continental Free Trade Area, Protocol on Trade in Goods, 2018, art 26 (j)). While this exceptions clause mirrors the exceptions clause set out in Article XX(j) of GATT, the emphasis on 'foodstuffs' with regard to products in short supply is unique to the AfCFTA. Indeed, Article XX(j) of GATT refers more broadly to products in short supply. Legally, the AfCFTA reference to foodstuffs in the context of products in short supply compared to the lack thereof in Article XX(j) of GATT is mostly insignificant, given that the broader language in Article XX(j) of GATT encompasses foodstuffs. Nevertheless, the direct reference to foodstuffs is significant in that it signals the importance that AfCFTA negotiators gave to food security considerations (Kuhlmann and Dall'Agnola 2023).

Ultimately, the extent to which member states can invoke Article 26 to justify measures that would otherwise be inconsistent with the Protocol on Trade in Goods depends on how an AfCFTA adjudicatory body would approach the issue (van der Ven and Signé 2021). Within the context of the WTO, exceptions have generally been difficult for member states to invoke successfully. In the context of interpreting 'products in general or local short supply', WTO adjudicatory bodies have examined the 'extent to which a particular product is "available" for purchase in a particular geographical area or market, and whether this is sufficient to meet demand in the relevant area or market' (WTO Appellate Body 2016). In doing so, the Appellate Body stressed not only the importance of looking at the domestic production of a product but also that 'due regard should be given to various factors, including the total quantity of imports that may be available to meet demand' (WTO Appellate Body 2016). Should the AfCFTA adjudicatory body adopt a similar interpretation, it would arguably set a high bar to invoking the 'food security' exception.

Agricultural disciplines that are absent in the African Continental Free Trade Area

In addition to identifying what is covered in the AfCFTA legal texts, it is equally important to identify what is not covered. In contrast to many regional trade agreements[3] and the WTO, as shown in Table 7.1, the AfCFTA includes neither a chapter on agriculture nor provisions directly relevant to agricultural production and food security, such as agricultural subsidy disciplines,

Table 7.1: Overview of agricultural provisions set out in the AfCFTA and the WTO

	AfCFTA	WTO
Agricultural chapter	✗	✓
Agricultural subsidies	✗	✓
Public stockholding	✗	✓
Special safeguard mechanism[4]	✗	✓

or provisions on public stockholding programmes that governments utilise to purchase, stockpile and distribute food when needed.

While agriculture was considered for inclusion at the AfCFTA drafting stage, it was later dropped given the lack of a compelling reason to have one (Desta 2023). Indeed, adopting disciplines on agricultural subsidies – the main objectives of the WTO Agreement on Agriculture – makes little sense if excessive subsidies are not a key problem. Most African countries lack the fiscal space to significantly subsidise their agricultural production. Indeed, the Agreement on Agriculture is especially concerned with generous amounts of domestic support in large agricultural producers such as China, the United States and the European Union. It aims to discipline these subsidies to curtail market distortions and price fluctuations that can destabilise the global agricultural market. Moreover, 44 out of 54 AfCFTA member states are also WTO members and therefore parties to a more extensive legal framework on agriculture in the WTO. Thus, the decision not to include a WTO-style protocol or annex on agriculture in the AfCFTA seems to have been a sensible one.

However, AfCFTA negotiators could have considered including additional food security provisions focused on enhancing regional cooperation to enhance food security and increase resilience.

This approach has been adopted by some of the RECs.[5] For example, the Treaty Establishing the East African Community (EAC) provides for initiating and maintaining 'strategic food reserves'. The Revised Treaty Establishing the Common Market for Eastern and Southern Africa (COMESA) and the Economic Community of West African States (ECOWAS) Revised Treaty allow for the conclusion of agreements on food security at the regional level (Treaty for the Establishment of the East African Community, 1999, art 110; Treaty Establishing the Common Market for Eastern and Southern Africa, 2012, art 131; Economic Community of West African States Revised Treaty, 1993, art 25).[6] COMESA also emphasises the importance of cooperation on the management of drought and desertification, whereas the EAC focuses on cooperation regarding the management of irrigation and water catchments, which can positively contribute towards achieving food security within these regions. Furthermore, COMESA sets out cooperation for the supply of staple foods including through investment, infrastructure provision, prevention of pre-and post-harvest losses, and an early-warning system to assess and supply information regarding food security, among others. The Intergovernmental Authority on Development (IGAD) also cooperates on supporting food security, through conducting research, supporting the development of sustainable agriculture in its member states, collaborating on the management of transboundary water and land governance, taking common measures to deal with transboundary pests (which can harm agricultural production) and supporting market access and policymaking in favour of resilient food systems (IFRAH IGAD Food Security Nutrition and Resilience Analysis Hub n.d.; IGAD 2024; IGAD n.d.).

Going beyond cooperation, the EAC also focuses on developing a mechanism for the exchange of information on demand and supply, surpluses and deficits, forecasting, and state of food nutrition, and to develop modalities to have timely information on market prices (Treaty for the Establishment of the East African Community, 1999, art 110). Moreover, it requires its member states to harmonise quality and standards of inputs and products, including on additives, as well as food supply, nutrition and food security policies and strategies, and to cooperate on the development of marine and inland aquaculture and fish farming (Treaty for the Establishment of the East African Community, 1999, art 110). EAC member states (officially referred to within the bloc as 'partner states') are encouraged to adopt good nutritional standards and the popularisation of indigenous foods (Treaty for the Establishment of the East African Community, 1999, art 118).

While many of the food security provisions in the examples provided from COMESA, EAC, ECOWAS and IGAD focus on cooperation and do not contain enforceable legal obligations, they provide insights into the types of food security provisions that the AfCFTA negotiators could have considered to strengthen the direct link between the AfCFTA and the continent's food security agenda.

Although African countries have continental frameworks for promoting food security (such as CAADP, Agenda 2063 and support for the African Union Commission's work on African agriculture), these are in some areas not as specific as the aforementioned REC agreements. This is particularly the case regarding the joint management of transboundary issues, such as the management of drought and desertification, water resources and pest control.

In sum, the AfCFTA's commitments on food security are limited, especially compared to relevant agriculture and food security provisions in the WTO and the RECs. While the absence of a WTO-style agriculture agreement is sensible, it is more difficult to see why REC-style cooperation provisions on food security did not find their way into the AfCFTA. Nonetheless, there are still significant ways in which the strategic implementation of the AfCFTA can prove essential to advancing food security in Africa, as we will see in the next section.

7.2 Implementing the African Continental Free Trade Area: removing non-tariff barriers critical for food security

As already noted, tariffs are generally low except for some peaks, such as in Somalia, where tariffs and other taxes on food are as high as 25 per cent (Mendez-Parra and Ayele 2023). Tariffs on agricultural inputs such as fertilisers and pesticides are also relatively high (Mendez-Parra and Ayele 2023). On the other hand, and as discussed in Chapter 5, results from several studies that model the AfCFTA's expected impact on food security emphasise that the greatest gains will come from tackling NTBs – defined in the AfCFTA as 'barriers that impede trade through mechanisms other than the imposition of tariffs' (Agreement Establishing the African Continental Free Trade Area, 2018, art 1(r)).

NTBs cover a diverse set of measures in terms of purpose, legal form and economic effect, and could include food safety regulations, elaborate testing requirements, rules of origin, and inefficient and costly border procedures.[7] By tackling NTBs, the AfCFTA can help galvanise intra-African trade in agri-food products, expand agricultural production, support food processing and value chain development, facilitate access to food, and develop more robust distribution networks. This will have knock-on effects in reducing Africa's relative dependence on food imports, while shielding the continent from severe supply-chain shocks.

This section assesses how the implementation of AfCFTA can reduce NTBs, with a focus on the Sanitary and Phytosanitary (SPS) Annex, the Technical Barriers to Trade (TBT) Annex, and various trade facilitation provisions. The assessment will be complemented in Section 7.3 with an overview of how implementation of some aspects of the Agreement on Trade in Services and the Phase II and Phase III protocols – including the AfCFTA's provisions on investment, digital trade, competition policies and intellectual property rights – will be instrumental to achieving food security in Africa.

Sanitary and phytosanitary measures in the African Continental Free Trade Area

SPS measures are critical for food security and public health. They ensure that minimum standards of safety are upheld, in order to protect human, plant or animal life or health. For example, food safety standards ensure that the food we eat do not contain harmful toxins, while governments and international organisations develop standards and guidelines to prevent spread of animal pests or diseases. At the same time, SPS measures can become significant barriers to trade, given their high compliance costs, which small producers and traders are often not able to meet. Indeed, the Food and Agriculture Organization of the United Nations has noted that domestic food prices in sub-Saharan Africa are 13 per cent higher, on average, as a result of SPS measures (Food and Agriculture Organization of the United Nations n.d.).

Annex 7 of the AfCFTA Agreement on Trade in Goods focuses on SPS measures (AfCFTA SPS Annex). From a food security perspective, four specific provisions set out in the AfCFTA SPS Annex will be particularly important: harmonisation, equivalence, cooperation and technical assistance. Harmonisation addresses the fragmentation of regulatory approaches by requiring member states to base their SPS measures on common standards. The AfCFTA SPS Annex provides that states 'shall cooperate in the development and harmonisation of sanitary or phytosanitary measures based on international standards, guidelines and recommendations taking into account the harmonisation of sanitary and phytosanitary measures at the regional level' (Agreement Establishing the African Continental Free Trade Area, 2018, art 8). Mirroring the WTO SPS Agreement, the AfCFTA SPS Annex refers to three international standard-setting bodies: the CODEX Alimentarius, the International Plant Protection Convention (IPPC) and the World Organization for Animal Health (formerly the International Office of Epizootics). Respectively, these organisations establish international rules for the use of toxic pesticides, invasive alien species, and veterinary medicines and animal diseases (Agreement Establishing the African Continental Free Trade Area, 2018, Annex 7, art 8).[8] Requiring AfCFTA member states to use international standards as the basis of their SPS measures reduces the compliance costs that traders face and therefore facilitates more food trade and increases consumer welfare (Mendez-Parra and Ayele 2023). Box 7.2 provides an example of the application of the harmonisation of maize standards in the EAC.

Another example concerns seed regulatory systems. Within the Southern African Development Community (SADC), different approaches to national seed regulation, including with regard to certification and quality control and quarantine measures for seed, made it difficult for seed to be traded within the region. Specifically, for a seed variety to be released in a SADC country, it would have to be tested for at least three seasons in different agro-ecological

Box 7.2: Harmonising maize standards in the EAC

Examples of food safety standards harmonisation can be found at the REC level. For instance, prior to 2005, EAC countries Kenya, Tanzania and Uganda applied different specifications for maize – including with regard to moisture content, aflatoxin levels, foreign matters and insect-damaged grains (see Table 7.2). In 2005, the EAC adopted harmonised standards for maize grains based on the Codex Alimentarius (and, in some cases, going beyond the standards set out in the codex). Doing so significantly facilitated maize trade within the EAC, as countries no longer had to ensure they complied with different SPS standards when trading with different countries.

Table 7.2: Comparison of maize SPS standards before and after EAC harmonisation and CODEX Alimentarius

	2003 (before EAC harmonisation)			After harmonisation		Codex Alimentarius (international standard)
	Kenya	Tanzania	Uganda	Grade 1	Grade 2	
Moisture content	13.5%	14%	13%	13.5%	13.5%	15%
Aflatoxin	10 ppb	10 ppb	10 ppb	10 ppb	10 ppb	Set by CODEX Commission
Foreign matter	1%	0.5%	1%	0.5%	1%	1.5%
Insect-damaged grains	3%	1%	2%	1%	2%	7%

Source: Reproduced and modified from John Keyser, Regional Quality Standards for Food Staples in Africa: Harmonization not Always Appropriate. July 2012. https://documents1.worldbank.org/curated/ar/357541468192844868/pdf/728540BRI0 Box30onal0Standards0FINAL.pdf

zones. The adoption of SADC's Harmonized Seed Regulatory System reduces the release procedure time, by allowing for any seed variety already approved in two SADC member states to be freely tradeable throughout the SADC region.[9] This facilitates access to, for instance, higher-yielding or drought-resistant seed varieties for farmers, to boost food production.

An alternative, less demanding approach to harmonisation is the concept of equivalence, which requires that an importing party accepts the SPS standards of another member state as equivalent to its own if the exporting party objectively demonstrates that the standards achieve the same level of SPS protection as the importing party – even if the requirements are not identical. For example, an importing country could consider equivalent an exporting country's approach to certifying organic agricultural products – allowing the product to be labelled in accordance with its own standards.[10]

Mirroring the WTO SPS Agreement, Article 7 of the AfCFTA SPS Annex requires that an importing party shall accept SPS measures of the exporting party as equivalent to its own if such equivalence can be objectively demonstrated. To advance food security within the African continent, it is recommended that the AfCFTA member states implement these provisions, including for food labelling, food safety practices and seed variety testing. The effective implementation of regionalisation provisions (Article 6) could also facilitate food security, given that, in the situation of a disease-outbreak, they allow for trade to continue from the country's disease-free zones.

A prerequisite to many of the provisions set out in the AfCFTA SPS Annex is that member states have in place a functioning SPS system. In many African countries, the SPS system is significantly underdeveloped. Member states should invest in and upgrade their SPS systems, including by taking advantage of the cooperation and technical assistance opportunities (Agreement Establishing the African Continental Free Trade Area, 2018, art 14), which include information sharing, developing and harmonising SPS measures at regional and continental level, developing infrastructure of testing laboratories, and developing centres of excellence could have significant gains to food security (Chinyamakobvu 2020).[11] Upgrading SPS systems will be critical to respond to environmental threats to crop production, including through pest disease outbreaks, which are expected to become more acute as a result of climate change.

The AfCFTA further contains various provisions that seek to streamline audit and verification (Article 9) as well as border check procedures related to import or export inspections and fees (Article 10), to ensure they are not more trade restrictive than necessary. These provisions can facilitate intra-African trade in food products, and are particularly important for agricultural and perishable goods, given their vulnerability to trade disruptions (Mendez-Parra and Ayele 2023). Other provisions of importance are those that seek to enhance transparency and the exchange of information (Article 11) and those that seek to ensure that traders have information as to the regulatory requirements.

Some progress has been made to harmonise food safety standards through the RECs, and to some extent across RECs through the African Organization for Standardisation (ARSO), of which 42 African countries are members. However, there is more to do to harmonise them at the continental level. This is where there is opportunity for effective implementation of AfCFTA SPS provisions (ARSO n.d.; Diop n.d., p.3; Economic Commission for Africa 2020, p.2).

Technical barriers to trade in the African Continental Free Trade Area

Similar to SPS measures, technical regulations and conformity assessment procedures can play a critical role in advancing legitimate policy objectives. At the same time, they have the potential to obstruct trade, including in agri-food products critical for food security. While technical regulations encompass most SPS measures, they are broader in scope and include regulations that go beyond protecting animal, plant or human life or health, and establish norms for packaging, technological specifications, labelling standards, the regulation of hazardous waste, and related issues. With regard to food security specifically, TBT standards can impact, for example, the way in which fish is caught, animals are treated, food is labelled, and what kind of technology is used in agricultural production.

Also, in line with the SPS Annex, the AfCFTA TBT provisions can facilitate intra-African trade in agri-food products by reducing the heterogeneity of technical and regulations and standards, which often create significant compliance burden and costs for exporters.[12] Annex 6 of the AfCFTA covers Technical Barriers to Trade. The Annex, which is based on the WTO TBT Agreement, aims to reduce NTBs by encouraging (1) cooperation in standards-setting, technical regulations, conformity assessments, accreditation and metrology; (2) the elimination of unnecessary and unjustifiable technical barriers to trade (Article 4); and (3) the promotion of mutual recognition of results in conformity assessments.[13] Similar to the AfCFTA SPS Annex, the implementation of these provisions could facilitate intra-African trade, including by streamlining standards and certification regarding warehouse storage, production, waste management, technology use, and other areas with implications for food security, impacting the quality, availability and affordability of food products.[14] In particular, the development of regional standards under ARSO and the AfCFTA Secretariat in areas relevant to the food supply chain could have important benefits to facilitating intra-African trade in agri-food products, as can the application of good regulatory practices (Article 7), transparency provisions (Article 11) and technical assistance and capacity-building (Article 12). A 2020 study by Economic Commission for Africa found that harmonisation of standards through the AfCFTA had potential to promote intra-African trade agri-food products and proposed a list of commodities for which this could be prioritised (Economic Commission for Africa 2020).

Trade facilitation in the African Continental Free Trade Area

Onerous document requirements and long export and import times are often a significant hindrance to trade across borders (Valensisi and Bacrot 2019). The following NTBs to trade are routinely experienced at many African borders: a lack of transparency in the rules and regulations, often resulting in discretionary decisions; delays and costs associated with border procedures; excessive bureaucracy; limited and uncoordinated working hours for customs personnel; the application of discriminatory taxes and other charges; cumbersome procedures for verifying containerised goods; and unpredictability in the requirements for product standards. These NTBs hinder intra-African food trade and increase the cost and time spent at borders. The latter can be particularly problematic for food products that are perishable.

Three AfCFTA Annexes to the Protocol on Trade in Goods contain provisions that seek to address high transaction costs of international trade. These are Annex 4 on Trade Facilitation – which is modelled on the 2013 WTO Trade Facilitation Agreement (TFA) – Annex 3 on Customs Cooperation, and Annex 8 on Transit – the latter two go beyond the TFA. These three annexes seek to streamline border crossing procedures, including by establishing a framework for the simplification and harmonisation of national customs legislation. They further require AfCFTA member states to establish modern data processing systems, use internationally accepted standards for customs automation, and set out provisions for communication and interpretation.

In particular, Annex 4 on Trade Facilitation contains provisions that are relevant to advancing food security objectives. These include publication of border requirements, enquiry points, advanced rulings, pre-arrival processing, transparent duties and charges, risk management, post-clearance audit, expedited shipments, and the exchange of information. It also contains a specific provision on perishable goods, defined as 'goods that rapidly decay due to their natural characteristics, in particular in the absence of appropriate storage conditions' (Agreement Establishing the African Continental Free Trade Area, 2018, Annex 4, art 1). Accordingly, the Trade Facilitation Annex recognises that perishable goods are more vulnerable than non-perishable goods to trade disruptions at international borders. For perishable goods, the Annex requires that AfCFTA member states release these goods within the shortest possible time under normal circumstances, and, exceptionally, outside the business hours of a customs authority (Agreement Establishing the African Continental Free Trade Area, 2018, Annex 4, art 15). It further requires that the member states ensure that perishable goods are given priority when scheduling examinations and that importing states arrange or allow for appropriate storage while perishable goods are being processed at the border.

These provisions in the Trade Facilitation Annex will enable member states to build upon good practices that are already emerging as a result of recent reforms. For instance, trade facilitation provisions at the REC level in the EAC require EAC member states to implement one-stop border posts at their common borders, to prevent dealing with customs at both the exporting country and importing country's border posts. According to a study by ODI, one-stop border posts in East Africa established through the support of TradeMark Africa have reduced total border crossing idle time between 62 per cent and 87 per cent, mainly due to time reductions in customs procedures. The study also found that this improvement in border crossing procedures had a direct impact on food prices: it reduced the price of maize by 9 to 12.3 per cent for maize sourced from Busia on the Uganda–Kenya border and 4.5 to 6.8 per cent for maize sourced from Taveta on the Kenya–Tanzania border. The study found similar results for rice, although with lower gains (Mendez-Parra and Ayele 2023).

In sum, implementing the provisions set out in the SPS, TBT and Trade Facilitation Annexes can play a critical role in achieving food security on the continent. A unique feature of the AfCFTA that can further facilitate the implementation of the AfCFTA NTB chapters is that under Annex 5 on Non-Tariff Barriers the AfCFTA provides for the establishment of a web-based NTB mechanism, which enables member states and economic operators (traders) to file complaints related to NTBs (see Box 7.3). The effective implementation and usage of this complaint mechanism could prove a game-changer to removing

NTBs on the continent. Indeed, specifically for situations involving perishable goods, it requires that, upon request by a member state, a specific complaint must be dealt with within 10 days and that, pending the final resolution, other interim solutions should be considered (Agreement Establishing the African Continental Free Trade Area, 2018, Annex 5, art 2.1.10–11).

Box 7.3: The NTB mechanism under the AfCFTA

Annex 5 of the Agreement Establishing the African Continental Free Trade Area (2018) uniquely provides for the establishment of a web-based NTB mechanism in which both member states and economic operators can file NTB-related complaints. The AfCFTA further requires member states to establish institutions to resolve NTBs through bilateral, pre-litigation dispute resolution. In doing so, it provides an opportunity for businesses located in any of the AfCFTA state parties to report NTBs encountered and set in motion a process for their resolution in a fast and cost-free manner.

Traders experiencing NTBs related to food security can submit complaints through this mechanism. Two web-based mechanisms already exist in Africa, covering four RECs: the Tripartite Free Trade Area (TFTA) NTB mechanism, hosted on tradebarriers.org, and the reporting and monitoring mechanism organized by Borderless Alliance, a private sector organisation in Ghana, hosted on tradebarrierswa.org. For a complaint to be submitted, member states must indicate whether the NTB concerns: (1) government measures that are trade and restrictive practices; (2) customs and administrative entry procedures; (3) technical barriers to trade; (4) sanitary and phytosanitary measures; (5) specific limitations; or (6) charges on imports.

Web-based NTB mechanisms are credited with increasing awareness of the challenges posed by NTBs. They provide a useful overview of the types of NTB that businesses in the region consider to be most problematic. NTB mechanisms have also been credited for their quick resolution of a significant percentage of complaints. For instance, the TFTA NTB mechanism resolved 597 complaints out of 663 that were registered. Among registered complaints were an SPS complaint by a Zambian company about having to submit a fumigation certificate for molasses; a TBT complaint by a Tanzanian company that Burundi had changed its labelling requirements after the company had submitted an export application; and a complaint related to specific limitations by Egypt about an import ban in Zimbabwe on soya bean oil packaged in bottles.

Source: www.tradebarriers.org; Agreement Establishing the African Continental Free Trade Area (2018), Annex 5.

7.3 The role of services, investment, digital trade, competition policy and intellectual property rights

Liberalising services and its implications for African food security

The Protocol on Trade in Services sets out principles for enhancing continental market access and service sector liberalisation (Tralac 2020). Some services, including financial, logistics, information and communication technologies, insurance, distribution and transport services, are intrinsically linked to food systems through agricultural production, distribution and trade, and through these channels to food security. Inter-African liberalisation of these services could attract investment and enhance competition with transformative impacts on agricultural production, value chains and food security.

Services negotiations under the AfCFTA follow an opt-in approach, which means that AfCFTA member states are only required to liberalise those services sectors in which they have made specific commitments. Five priority sectors have been adopted for services liberalisation: financial services, transport, communication services, business services, and tourism and travel. Commitments can be made for each of these sectors for different modes of supply, as is elaborated upon in the context of agricultural production and food security in Box 7.4.

Box 7.4: Modes of services commitments relevant to agricultural service/food security

Following the WTO General Agreement on Trade in Services, market access commitments in services are scheduled per service 'mode' of supply. Specifically, AfCFTA member states can use four different services modes in their schedules:

- **Mode 1: Cross-border supply**, e.g. a farmer from country A gets crop insurance from a company based in country B.

- **Mode 2: Consumption abroad**, e.g. a consumer from country A travels to country B to access repair services to fix a broken smart technology weather app on a phone.

- **Mode 3: Commercial presence/foreign direct investment**, e.g. an agricultural drip irrigation technology company from country A opens a branch office in country B to install the technological equipment.

- **Mode 4: Presence of natural persons**, e.g. a veterinary official from country A travels to country B to check on a farmer's cattle.

Among the five services that are prioritised, commitments in financial services, communication services and transport will have most impact for African food security. Financial services are integral to the development of the agricultural sector. This includes financial services such as credit, deposits, payment, insurance and risk management services (Chandra and Kinasih 2012). As was seen in Chapter 3, African farmers have very low levels of financial inclusion. Only 17 per cent use a formal financial institution for savings and only 10 per cent for borrowing (Madden 2020). Increasing access to financial services, including through the liberalisation of financial services on the continent, could increase the uptake of financial services utilisation, including in the agricultural sector (Madden 2020). It would similarly be critical to increase the digitisation of agribusiness payments, for which implementing the Protocol on Digital Trade, including provisions on financial technology, discussed in more detail below, will be critical. It is, however, not automatic that the liberalisation of financial services will trickle down to smallholder farmers – as this is not necessarily guaranteed (Dube 2012).

Logistics services, including transport, and information and communication technology are critical to reduce the costs and uncertainty in agricultural trade. A systematic review of the literature published in 2016 found that improving rural road infrastructure leads to higher agricultural production and, as a result, higher incomes (Hine et al. 2016). More recent evidence mostly corroborates these findings.[15] Similarly, studies have found a positive correlation between reforms in distribution services and the transformation of food systems for farmers, increased food security, and decrease in rural poverty (Chandra and Kinasih 2012). Well-functioning logistics services can also shorten supply chains and improve the availability, quality, safety, price and variety of food products.

The liberalisation of services under the AfCFTA can positively impact food security in Africa, given that it will open these sectors to competition making provision of these services more effective. At present, many services sectors on the African continent are underdeveloped, highly regulated, and mostly monopolised by government parastatals, resulting in high costs and inefficiencies (Dube 2012). AfCFTA member states should aspire to make ambitious commitments in these areas, subject to country-specific contexts.

Leveraging the Protocol on Investment to advance food security

As discussed in Chapter 3, public and private investment in equipment and infrastructure including irrigation systems, storage facilities and the mechanisation of production is far from optimal (Petrack n.d.). According to a study by McKinsey & Company, countries south of the Sahara will require investment of US$8 billion for improved storage, and US$65 billion for irrigation, for the continent to achieve its agricultural potential (Goedde, Ooko-Ombaka and Pais 2019). To enhance production efficiencies, farmers require eight times more fertiliser and six times more quality seeds than current levels (Petrack n.d.). Investment is also required for transport infrastructure and, more generally, for cross-border value chain development (African Union Commission and OECD 2022).

The Protocol on Investment promotes intra-African investments in these and other areas relevant to food security, as it contains provisions to reduce the risks associated with cross-border investment, along with provisions to promote and facilitate investment, while balancing this with sustainability considerations and carve-outs. For instance, the Protocol on Investment includes provisions that protect investors from discriminatory treatment vis-à-vis other African investors (Article 11 on national treatment and Article 13 on most-favoured nation treatment); provisions that provide investors and their investments physical protection and security (Article 16); provisions that protect investors from expropriation (Article 17); and provisions that enable the transfer of profits and other returns from investment (Article 19). For each of these provisions, the Protocol on Investment has crafted exceptions and carve-outs related to sustainability concerns.

In addition, investor protection provisions are balanced with provisions that establish investor obligations, including compliance with national and international law, business ethics, human rights and labour standards, environmental protection, indigenous peoples and local communities; sociopolitical obligations; anti-corruption; corporate social responsibility; and investor liability. These provisions ensure that the investments that fall within the scope of the Protocol on Investment respect basic human rights and sustainability requirements.

Other provisions set out in the Protocol on Investment that could contribute to facilitating intra-African investments, including in agriculture, are those on investment promotion and facilitation (Chidede 2019). With regard to investment promotion, the Protocol on Investment enables member states to adopt incentives to 'encourage preferential markets schemes and specific investors within the region' as well as incentives to provide for technical assistance and technology transfer requirements (Protocol on Investment to the Agreement Establishing the African Continental Free Trade Area, 2023, art 24). At the same time, investment facilitation provisions could address issues such as excessive bureaucracy, lack of transparency about investment-related information, corruption, and lack of coordination of relevant institutions, which are key issues that hinder intra-African investment flows.

In sum, the Protocol on Investment could play a catalytic role in enhancing intra-African investment, including with regard to the agricultural sector. The benefits, however, will not be automatic, and require adopting a proactive and strategic approach to enhancing intra-African investment. In addition, the continent will still rely heavily on investment from outside Africa, especially in areas related to high-yielding seed varieties or mechanisation of agriculture. These investors are not directly covered by the Protocol on Investment but in some cases covered by bilateral investment treaties between individual African countries and foreign partners.

Scaling up digital agricultural solutions through the Protocol on Digital Trade

The Protocol on Digital Trade can help to create a digital enabling environment that can boost the uptake of digital technologies that are critical to increasing agricultural yields and enhancing food preservation. For instance, this includes mobile phone applications that can used to buy and sell seeds and inputs, or that set out early-warning systems regarding weather events; digital technologies that enable up-to-date tracking of commodities that are being transported to markets; access to real-time product prices; automated drip-irrigation technologies; or optimise crop pests/disease mitigation strategies (AUDA-NEPAD African Union Development Agency 2021).

Digital agricultural solutions being used in Africa include the Hello Tractor app, also known as the 'Uber for tractors', which enables farmers in 13 countries, including Nigeria, Kenya and Tanzania, to rent tractors and equipment at affordable rates (Bhalla 2021), or DigiFarm in Kenya, which serves as a one-stop shop that enables farmers to bypass middlemen to access low-cost seeds and fertilisers, loans, insurance and so on, and many others. Yet scaling these solutions remains a challenge.

The implementation of the provisions set out in the AfCFTA Protocol on Digital Trade can further facilitate the use of digital solutions for smart agriculture across borders, with positive impacts along the food system's value chain. In particular, the protocol requires member states to refrain from imposing customs duties on digital products that are transmitted electronically, introduces several trade facilitation provisions, including digital contracts, electronic invoicing, digital payments, and paperless trading and last-mile delivery. It also requires that the member states allow for the cross-border transfer of data (Article 20), including personal data, and to refrain from requiring to use or locate computing facilities in a member's territory as a condition for engaging in digital trade, both of which will be critical to scaling up smart agricultural solutions and applications. However, this obligation is subject to exceptions, to achieve a legitimate public policy objective or protect essential security interests. Provisions that seek to enhance data innovation, including by collaborating on various data-sharing projects and sharing best practices, could also catalyse digital innovation across the continent, including smart agriculture applications.

Uniquely, Part VI of the Protocol contains provisions relevant to digital trade inclusion, which seeks to promote the inclusion and participation of women, youth, indigenous people and rural and local communities in digital trade. This is also directly relevant to enhancing the uptake of digitalisation in agricultural production in rural areas and as regards agricultural activities predominantly carried out by women. It also includes provisions directed at the inclusion of micro, small and medium-sized enterprises (MSMEs) by focusing on financing and skills development and provisions that seek to facilitate the adoption, development and collaboration in relation to emerging and advanced technologies. These provisions, if applied strategically vis-à-vis MSMEs active in agriculture and agricultural technologies that can enhance production efficiency, could be critical in facilitating Africa's transition to becoming a food-secure continent.

Mainstreaming women and youth in trade activities

The provisions in the Protocol on Digital Trade are reinforced in the Protocol on Women and Youth, which aims to enhance intra-African trade participation among women and youth entrepreneurs and business owners by addressing a number of challenges that women have historically faced such as access to trade finance, participation in trade policymaking, support to enhance export capacity, and a range of trade facilitation measures that have been gender-blind.

Addressing anticompetitive behaviour in food sectors through the Protocol on Competition Policy

Both globally and within the African continent, there is an increasing economic concentration in the production and trading of agriculture and food products (Roberts 2023). While such consolidation enhances vertical and horizontal integration, providing, for instance, farmers with bundles of goods and services across food systems, it also means that large firms can exert market power to raise prices to consumers, while restricting the participation of smaller players (Roberts 2023). A study of maize and soya bean market dynamics in Eastern and Southern Africa identified price fluctuations that can be traced back to excessive mark-up at trader level, government intervention, and poor options with regard to storage and logistics (Roberts 2023). These findings highlighted that smallholder farmers in a subset of East and Southern African countries received unfairly low prices for their maize products, while fish and poultry farmers, who buy soya beans as feed, were not able to compete with imported frozen fish and chicken (Roberts 2023).

The implementation of the AfCFTA Protocol on Competition Policy could play an important role in addressing anticompetitive behaviour in the food sector. While the Protocol on Competition Policy does not directly refer to food security or agriculture, it highlights among its objectives the promotion of economic integration and sustainable development in the AfCFTA market (Article 2). It notes practices such as abuse of dominant position in the market, mergers or acquisitions that restrict or prevent competition, and abuse of economic dependence (Article 5). Importantly, it establishes a continental authority, with an investigative body, to administer and enforce the provisions set out in the Protocol on Competition Policy, with the ability to impose sanctions where it finds anticompetitive behaviour to exist.

By addressing and sanctioning anticompetitive practices, the Protocol on Competition Policy seeks to ensure that the benefits associated with the creation of the AfCFTA would not be undermined by anticompetitive behaviour, including in the food industry. In practice, the real impact of the Protocol on Competition Policy will depend on (1) whether a member state has a functioning competition authority in place and (2) the extent to which member states will be using the continental authority and investigative body on competition.

Protections through the Protocol on Intellectual Property Rights

The AfCFTA Protocol on Intellectual Property Rights will apply to all categories of intellectual property right, including plant varieties, geographical indications, genetic resources and traditional knowledge. The protocol can have important implications for food security. As part of its 'guiding principles', it includes the 'promotion of the public interest in sectors of vital importance to socio-economic and technological development including ... agriculture, food security, and nutrition'.[16] Similarly, under the heading 'areas for cooperation' it includes a direct reference to 'the use of flexibilities under international instruments for the protection of food security, agriculture, and nutrition' (Protocol to the Agreement Establishing the African Continental Free Trade Area on Intellectual Property Rights, 2023, art 23).

Various provisions will have particular relevance for agricultural production and food security. Nearly half of all African countries already have an intellectual property system in place for seeds, most of them following the model of the 1991 Convention of the International Union for the Protection of New Varieties of Plants. Article 8 of the AfCFTA Protocol on Intellectual Property Rights on the protection of new plant varieties will put in place a *sui generis* or unique system that includes farmers' rights, plant breeders' rights and rules on access and benefit sharing across the continent. Exactly what this would look like will be further developed as part of an annex that will be added to the protocol. Putting such protections in place can incentivise investment in innovation in the development of new, higher-yielding or drought- and heat-tolerant plant varieties. In developing the annex on the protection of new plant varieties, it would be important to ensure that incentives for investment in new varieties are balanced by adequate access and benefit sharing provisions, to ensure that farmers are not prevented from using new plant varieties.

The AfCFTA further provides protection for geographical indications (GIs), including for agricultural products that are connected to geographic areas, which can enhance food security by preserving and promoting traditional products both in local and international markets. Specifically, the protocol aims to provide protection of geographical indications and establish a database and information portal of registered geographical indications – with an annex setting out additional obligations to be further developed (Protocol to the Agreement Establishing the African Continental Free Trade Area on Intellectual Property Rights, 2023, art 9). There are a large number of traditional products that can benefit from GI protection, including penja pepper and rooibos tea in South Africa, Casamance honey in Senegal, teff from Ethiopia, Maferinyah pineapple from Guinea, and Bondoukou Kponan yarn from Côte d'Ivoire (African Union n.d.). Including geographical indications in the AfCFTA enhances global recognition of the protected products.

The protocol further includes a provision protecting genetic resources and traditional knowledge. Each provision requires that applications for an intellectual property right that is drawn from genetic resources or traditional knowledge must provide various types of information, including disclosure of source, proof of prior informed consent, and proof of fair and equitable benefit sharing. More generally, the protocol sets out a number of other requirements, and requires that relevant African and international instruments can be used to further develop rules on prior informed consent and so on. Additional obligations will be developed and annexed to the protocol on traditional knowledge, traditional cultural expression, folklore and genetic resources. These provisions could also enhance food security by protecting traditional knowledge, which can enable indigenous communities to benefit economically.

While various intellectual property rights frameworks have already been adopted within the African continent, the benefit of the provisions of the protocol is that it aims to harmonise existing approaches, thereby creating less fragmentation. This will be crucial for developing intra-African value chains. To maximise benefits for agricultural production and food security, it is important that the annexes that are yet to be developed in the area of protection of new plant varieties, GIs, and traditional knowledge, traditional cultural expression, folklore and genetic resources strike the right balance between protecting intellectual property and incentivising much-needed innovation while ensuring fair and affordable access to these innovations – keeping in mind the interests of millions of smallholder farmers in the continent. This will help to ensure a balance between the costs and benefits of intellectual property protections (ECA, AU and AfDB 2017, pp.14–153; ECA, AU and AfDB 2016; United Nations Economic Commission for Africa et al. 2019, pp.103–31).

7.4 Environmental provisions in the African Continental Free Trade Area

As we argue throughout this book, there are strong links between agricultural production and climate change/environmental degradation, with the former contributing to the latter, and the latter enhancing risks. This section looks at ways in which the AfCFTA legal instruments can help to mitigate these risks.

As we saw earlier in relation to food security, the AfCFTA texts contain only minimal references to the environment. The preamble to the Agreement reaffirms the right of member states to regulate within their territories on climate and sustainable development matters. The preamble to the Protocol on Trade in Services also recognises the right of member states to adopt measures by introducing services regulations to meet legitimate policy objectives, including on sustainable development. However, as noted earlier in the context of food security, preambular citations do not amount to legally binding obligations.

As we also saw in relation to food security, the Protocol on Trade in Goods contains exception clauses for situations where AfCFTA member states parties adopt environmental sustainability measures. Specifically, the protocol provides that, under certain circumstances, member states may adopt environmental sustainability measures that are inconsistent with the Protocol on Trade in Goods, including if these measures are 'necessary to protect human, animal or plant life or health' or 'relat[e] to the conservation of exhaustible natural resources if such measures are made effective in conjunction with restrictions on domestic production or consumption' (Agreement Establishing the African Continental Free Trade Area, 2018, art 26). Provided certain conditions are met, the adoption of measures that are necessary to protect human, animal or plant life or health could justify violations of provisions in the Protocol on Trade in Goods.

As we have already seen, the Protocol on Intellectual Property Rights contains a few references to the environment. It includes as one of its guiding principles the 'facilitation of access to clean and efficient energy, as well as promote just and fair energy transition and environmental sustainability' (Protocol on Investment to the Agreement Establishing the African Continental Free Trade Area, 2023, art 4). It further encourages AfCFTA member states to register marks, patents and industrial designs for environmentally friendly goods and services, designs and innovations.

More notably, the Protocol on Investment stands out for extensive environmental references. For instance, it recognises that an investment's impact can be a factor to consider in establishing whether, for discrimination purposes, two investments are made in 'like circumstances' (Protocol on Investment to the Agreement Establishing the African Continental Free Trade Area, 2023, art 11). It further includes exceptions to violations of national treatment, most-favoured nation and expropriation provisions for regulatory measures designed to protect the environment. The Protocol on Investment goes further to establish minimum standards on the environment, noting that AfCFTA member states must ensure high levels of environmental standards and shall not encourage investments by relaxing compliance with environmental standards (Protocol on Investment to the Agreement Establishing the African Continental Free Trade Area, 2023, art 22). It requires investors and their investments to respect and protect the environment while carrying out their business activities, including respecting the right to a clean and sustainable environment; complying with the principles of prevention and precaution to anticipate significant harm to the environment; carrying out environmental impact assessments; and mitigating and restoring any environmental harm that companies have caused (Protocol on Investment to the Agreement Establishing the African Continental Free Trade Area, 2023, art 30). The protocol further establishes corporate social responsibility for investors, with various references to environmental protection (Protocol on Investment to the Agreement Establishing the African Continental Free Trade Area, 2023, art 34). These environmental references are extensive compared to other investment agreements.

In sum, except for the Protocol on Investment, the AfCFTA legal framework does not contain prominent provisions on climate and the environment. However, implementation of the SPS and TBT annexes, removing and reducing tariffs and NTBs on environmental goods and services and application of the relevant measures in the Protocols on Investment, Competition Policy, Intellectual Property Rights and Digital Trade can lead to progress on various sustainability matters.[17]

Summary

Direct references to food security in the AfCFTA legal instruments are limited. Unlike the WTO Agreement on Agriculture, or the food security cooperation approaches adopted in the RECs, the AfCFTA legal texts neither contain elaborate provisions to discipline agricultural subsidies – for good reasons – nor do they contain provisions to enhance cooperation on food security at the continental level. However, the absence of provisions on regional food security modelled after the RECs can be seen as missed opportunity. It is for this reason that some scholars have advocated the need for the development of a Protocol on Food Security.[18]

Despite the limited references to food security in the AfCFTA legal texts, implementation of its protocols and annexes can also have a positive effect on agricultural production, with significant benefits for food security. This is especially the case with AfCFTA provisions that aim to ensure SPS and TBT compliance, promote trade facilitation provisions that seek to streamline border processes, which is critical for perishable goods, and more broadly discipline NTBs. The creation of a web portal where traders and governments can submit complaints about NTBs is an important initiative that could further facilitate the implementation of these provisions. The protocols on services, investment, digital trade, competition policy and intellectual property rights, if implemented effectively, could boost intra-African value chains in agriculture and agribusinesses, enhance efficiency and lower prices.

But none of this will happen automatically. First, it will require that AfCFTA member states proactively apply the measures and start trading under the AfCFTA. Despite the AfCFTA's official launch in January 2021, trading under its legal instruments is, at the time of writing, yet to commence. Here, the Guided Trade Initiative, launched in October 2022, which seeks to kick-start trade under the AfCFTA should quickly transition to comprehensive implementation (Rao 2022).

Notes

[1] Although the simulations in Chapter 6 find that the impacts of the AfCFTA on intra-African food trade will be modest, these impacts could be larger in times of food shortages (particularly where these affect some countries more than others). This is because the simulations in Chapter 6 are based on the state of Africa's economies from 2017 to 2019. However, if there were a shortfall of food from other sources, the number of people needing to source food from elsewhere in Africa (and, as a result, the number who would find it more affordable to do so as a result of tariff reductions) could increase. Even though Africa is a net food importer, intra-African trade could still be important for addressing food shortages in a time of crisis if rising prices of foods that Africa consumes made it more attractive for African farmers to supply the continent's food market instead of focusing on exporting elsewhere and/or producing non-food crops.

[2] This chapter uses the term 'member state(s)' to refer to African Union members that have signed and deposited instruments of ratification for the AfCFTA. For consistency throughout the book, it uses 'member states' even though the official term used is 'State Parties'.

[3] See, e.g., OECD (2015).

[4] The AfCFTA does include a global safeguards provision (Article 18 of the Protocol on Trade in Goods), mirroring Article XIX of GATT 1994 and the WTO Agreement on Safeguards. It also includes a preferential safeguards provision (Article 19 of the Protocol on Trade in Goods), which allows state parties to apply safeguard measures in specific situations. These provisions can be used to protect farmers from excessive price volatilities, including agriculture commodities that are included in the AfCFTA member states' tariff schedules. Given that intra-African trade is mainly concentrated within Africa's five main regions, and the fact that African countries rarely resort to using safeguards, it is unlikely that these provisions will be invoked by African countries in the context of agriculture.

[5] See also Kuhlmann and Dall'Agnola (2023).

[6] If that means the possibility of member states maintaining regional food reserves, it might raise specific concerns under the WTO disciplines, given that WTO law does not recognise the concept of regional food security.

[7] While governments may – and should – adopt non-tariff measures to pursue various policy objectives, such as protecting human health and safety, plant and animal health, or environmental concerns, these measures become NTBs where they constitute unjustifiable trade restraints.

[8] Under the SPS Annex, AfCFTA member states may also introduce higher SPS standards, but these could be subject to a scientific justification or the result of a risk assessment (Agreement Establishing the African Continental Free Trade Area, 2018, Annex 7, art 8).

[9] Feed the Future (n.d.).

[10] Japan has adopted this approach vis-à-vis organic agricultural products from the US. See Bellmann and van der Ven (2020).

[11] See also van der Ven and Signé (2021).

[12] See also van der Ven and Signé (2021).

[13] Agreement Establishing the African Continental Free Trade Area (2018, Article 4 and Annex 6).

[14] Kuhlmann and Dall'Agnola (2023).

[15] An exception is some evidence from India. Relevant literature published since the systematic review identified by the author through a rapid review includes Anega and Alemu (2023), Asher and Novosad (2020), Berg, Blankespoor and Selod (2019), Gennadevich Bryzhko and Viktorovich Bryzhko (2019), Hine et al. (2019), Nakamura, Bundervoet and Nuru (2020), Saifullah Kamaludin and Mariatul Qibthiyyah (2022), Takada et al. (2021) and World Bank (2023).

[16] Protocol to the Agreement Establishing the African Continental Free Trade Area on Intellectual Property Rights (2023, Article 4 (d)).

[17] For further reading, see van der Ven and Signé (2021).

[18] See Kuhlmann and Dall'Agnola (2023).

References

African Union (n.d.) 'Continental Strategy for Geographical Indications in Africa 2018–2023'. https://perma.cc/UN9M-2HWF

African Union Commission; and OECD (2022) *Africa's Development Dynamics 2022: Regional Value Chains for a Sustainable Recovery*, Paris: OECD Publishing. https://doi.org/10.1787/2e3b97fd-en

Agreement Establishing the African Continental Free Trade Area, 2018.

Agreement Establishing the African Continental Free Trade Area, Protocol on Trade in Goods, 2018.

Agreement on Safeguards, 1994.

Appellate Body (2016) 'India – Certain Measures Relating to Solar Cells and Solar Modules. AB-2016-3', World Trade Organization. https://docs.wto.org/dol2fe/Pages/SS/directdoc.aspx?filename=q:/WT/DS/456ABR.pdf&Open=True

Asher, Samuel; and Novosad, Paul (2020) 'Rural Roads and Local Economic Development', *American Economic Review*, vol. 110, no. 3, pp.797–823. https://doi.org/10.1257/aer.20180268AUDA-NEPAD African Union Development Agency (2021) 'Smart Agriculture through Mobile Technologies in Africa', AUDA-NEPAD African Union Development Agency, 25 January. https://perma.cc/QB58-H2AQ

Bellmann, Christophe; and van der Ven, Colette (2020) 'Greening Regional Trade Agreements on Non-tariff Measures through Technical Barriers to Trade and Regulatory Co-operation', OECD Trade and Environment Working Papers, No. 2020/04. https://doi.org/10.1787/dfc41618-en

Berg, Claudia N.; Blankespoor, Brian; and Selod, Harris (2019) *Roads and Rural Development in Sub-Saharan Africa*, Washington, DC: World Bank. https://perma.cc/DB4N-B9JH

Bhalla, Nita (2021) 'Feature-Africa's Farmers Click with Digital Tools to Boost Crops', *Reuters*, 14 October. https://www.reuters.com/article/africa-tech-farming-idUSL4N2QU29J

Bryzhko, Viktor G.; and Bryzhko, Ilya V. (2019) 'Comprehensive Assessment of the Impact of Road Infrastructure Development in a Rural Municipal Area (Russia)', *Revista ESPACIOS*, vol. 40, no. 37. https://perma.cc/BT8J-P8H2

Chandra, Alexander C.; and Kinasih, Herjuno N. (2012) 'Services Trade Liberalization and Food Security: Exploring the Links in the Association of Southeast Asian Nations (ASEAN)', Winnipeg: International Institute for Sustainable Development. https://perma.cc/7FVE-6PME

Chidede, Talkmore (2019) 'How Can the AfCFTA Investment Protocol Advance the Realization of the AfCFTA Objectives?', *tralac*, 12 June. https://www.tralac.org/blog/article/14065-how-can-the-afcfta-investment-protocol-advance-the-realisation-of-the-afcfta-objectives.html

Chinyamakobvu, Oswald (2020) 'Update on the African Continental Free Trade Area (AfCFTA)', IPPC Regional Workshop, 8–11 September.

Desta, Melaku (2023) 'Intervention by Melaku Desta'. *Workshop on African trade and food security*, London School of Economics and Political

Science, June.Diop, Aissatou (n.d.) 'Standards in the Context of the African Continental Free Trade Area', Geneva: International Trade Centre. https://perma.cc/8PRD-9D7U

Dube, Memory (2012) 'Leveraging Services Trade Liberalization for Enhanced Food Security in the Southern Africa Development Community', Winnipeg: International Institute for Sustainable Development. https://perma.cc/8TFS-QJK4

ECA, AU and AfDB (2017) 'Assessing Regional Integration in Africa VIII: Bringing the Continental Free Trade Area About', Addis Ababa: United Nations Economic Commission for Africa, African Union and African Development Bank. https://doi.org/10.18356/06269c87-en

ECA, AU and AfDB (2016) 'Assessing Regional Integration in Africa VII: Innovation, Competitiveness and Regional Integration', Addis Ababa: United Nations Economic Commission for Africa, African Union Commission and African Development Bank. https://doi.org/10.18356/73293cf1-en

Economic Commission for Africa (2020) *Identifying Priority Products and Value Chains for Standards Harmonization in Africa*, Addis Ababa: ECA. https://perma.cc/U8FF-YNJD

Feed the Future (n.d.) 'The Southern African Development Community (SADC) Harmonized Seed Regulatory Systems'. https://perma.cc/B2X3-DF5C

Food and Agriculture Organization (n.d.) 'Trade and Food Safety Standards. African Free Trade and Food Safety', Rome: FAO. https://perma.cc/R4QL-2JP8

Goedde, Lutz; Ooko-Ombaka, Akinwumi; and Pais, Gillian (2019) 'Winning in Africa's Agricultural Market', McKinsey & Company. https://perma.cc/787G-K4CC

Hine John; Abedin Masam; Stevens Richard J; Airey Tony; and Anderson Tamala (2016) 'Does the Extension of the Rural Road Network Have a Positive Impact on Poverty Reduction and Resilience for the Rural Areas Served? If so How, and if not Why Not? A Systematic Review', London: EPPI-Centre, Social Science Research Unit, UCL Institute of Education, University College London. https://perma.cc/9TQ9-2QSR

Hine, John; Sasidharan, Manu; Eskandari Torbaghan, Mehran, Burrow, Michael; and Usman, Kristianto (2019) 'Evidence of the Impact of Rural Road Investment on Poverty Reduction and Economic Development', Brighton: Institute of Development Studies. https://perma.cc/JLM9-Y363

IFRAH IGAD Food Security Nutrition and Resilience Analysis Hub (n.d.) 'ICPAC IGAD Climate Prediction and Applications Centre'. https://perma.cc/3M2K-CCVL

IGAD (2024) 'Launching of Food Systems Resilience Program for Eastern and Southern Africa IGAD Peace, Prosperity and Regional Integration'. https://igad.int/launching-of-food-systems-resilience-program-for -eastern-and-southern-africa/

IGAD (n.d.) 'Natural Resources Management', IGAD Peace, Prosperity and Regional Integration. https://perma.cc/M6WU-5CKH

Kamaludin, Ahmad Saifullah; and Qibthiyyah, Riatu Mariatul (2022) 'Village Road Quality and Accessibility on Transforming Rural Development', AGRARIS: Journal of Agribusiness and Rural Development Research, vol. 8, no. 2, pp.160–80. https://doi.org/10.18196/agraris.v8i2.13618

Kuhlmann, Katrin; and Dall'Agnola, Giacomo (2023) 'Breaking New Ground on Food Security: Africa's Potential to Reframe Global Rules through the AfCFTA', Georgetown Law Working Paper Draft. https://doi.org/10.2139/ssrn.4943557

Madden, Payce (2020) 'Figures of the Week: Financial Inclusion of Farmers in Africa', Brookings. https://perma.cc/UP77-BSN9

Mendez-Parra, Maximiliano; and Ayele, Yohannes (2023) 'How African Integration Can Help Achieve Food Security', ODI, 16 October. https://odi.org/en/insights/how-african-integration-can-help-to -achieve-food-security/

Nakamura, Shohei; Bundervoet, Tom; and Nuru, Mohammed (2020) 'Rural Roads, Poverty, and Resilience: Evidence from Ethiopia', The Journal of Development Studies, vol. 56, no. 10, pp.1838–55. https://doi.org/10.1080/00220388.2020.1736282

Naod Mekonnen, Anega; and Alemu, Bamlaku (2023) 'The Effect of Rural Roads on Consumption in Ethiopia', Journal of Economics and Development, vol. 25, no. 3, pp.186–204. https://doi.org/10.1108/jed-10-2022-0213

OECD (2015) 'Regional Trade Agreements and Agriculture', OECD Food, Agriculture and Fisheries Papers, no. 79, Paris: OECD Publishing. https://perma.cc/ZUJ4-SVYL

Petrack, Shira (n.d.) 'Invest in African Agriculture', Empower Africa. https://perma.cc/3HW7-W7EV

Protocol to the Agreement Establishing the African Continental Free Trade Area on Intellectual Property Rights, 2023.

Protocol on Investment to the Agreement Establishing the African Continental Free Trade Area, 2023.

Rao, Pavithra (2022) 'AfCFTA's Guided Trade Initiative Takes Off, Set to Ease and Boost Intra-African Trade', Africa Renewal, 12 October. https://perma.cc/64EM-42CL

Roberts, Simon (2023) 'Competition, Trade, and Sustainability in Agriculture and Food Markets in Africa', *Oxford Review of Economic Policy*, vol. 39, no. 1, pp.147–61. https://doi.org/10.1093/oxrep/grac041

Takada, Shin; Morikawa, So; Idei, Rika; Kato, Hironori (2021) 'Impacts of Improvements in Rural Roads on Household Income through the Enhancement of Market Accessibility in Rural Areas of Cambodia', *Transportation*, vol. 48, no. 5, pp.2271–95. https://doi.org/10.1007/s11116-020-10150-8

Tralac (2020) 'Trade in Services Negotiations under the AfCFTA'. https://www.tralac.org/documents/resources/faqs/3190-trade-in-services -negotiations-under-the-afcfta-q-a-march-2020/file.html

United Nations Economic Commission for Africa, AU, AfDB and UNCTAD (2021) 'Assessing Regional Integration in Africa IX: Next Steps for the African Continental Free Trade Area', Addis Ababa: United Nations Economic Commission for Africa, African Union Commission, African Development Bank and United Nations Conference on Trade and Development. https://perma.cc/62D8-4ZHE

Valensisi, Giovanni; and Bacrot, Celine (2019) 'Harnessing Trade Facilitation for Regional Integration', UNCTAD, 4 December. https://perma.cc/YB8V-9QM6

Van der Ven, Colette; and Signé, Landry (2021) 'Greening the AfCFTA: It's Not Too Late', Brookings Institute, September. https://perma.cc/62BR-WBDJ

World Bank (2023) 'How Are Roads Changing Lives in Madagascar?', The World Bank. https://perma.cc/UW7J-72M9

8. Africa's bilateral food trade

Vinaye Dey Ancharaz

This chapter on Africa's bilateral food trade (i.e. trade with non-African partner countries) turns the spotlight onto the changing pattern of trade with
traditional partners and the growing relationship with emerging partners. It
complements the discussion in Chapter 2 (on Africa's global trade flows and
its decomposition into agricultural trade flows, food trade flows and agricultural inputs trade flows), in Chapter 5 (on formal and informal intra-African food trade) and in Chapter 9 (on the World Trade Organization (WTO)
legal framework as it pertains to food trade and food security). In this chapter, bilateral food trade flows and the trade regimes underpinning them are
brought into sharper focus. The chapter begins by examining the changing
patterns in Africa's food and agricultural imports and exports before going on
to discuss the trade policy aspects of these flows.

8.1 Traditional and emerging partners

The enduring story of Africa's bilateral food trade is the changing shares of the
partners in both exports and imports since the turn of the century. While
the value of exports and imports in current prices have increased (Figures 8.1
and 8.2), the proportion that is with countries of the Global South and that is
from intra-African trade itself have grown significantly.

Among the traditional partners, the European Union (EU) remains the
principal market for sourcing food imports and destination for food exports.
But the EU's shares are declining. In 2017–2021, the European bloc received
27.7 per cent of Africa's food exports, compared to 33.2 per cent a decade
earlier. On the import side, the decline was from 23.3 per cent to 21.5 per cent
from 2007–2011 compared to 2017–2021. Even so, the EU remains the most
important partner for Africa's food trade alongside the continent's trade with
itself. The same trend in lower shares in the value of exports and imports is
also evident in the case of the United Kingdom (UK). Imports from the

How to cite this book chapter:

Ancharaz, Vinaye Dey (2025) 'Africa's bilateral food trade', in: Luke, David (ed)
How Africa Eats: Trade, Food Security and Climate Risks, London: LSE Press,
pp. 187–211. https://doi.org/10.31389/lsepress.hae.h License: CC-BY-NC 4.0

Figure 8.1: Africa's exports by destination, basic food (US$ billion, current prices), 2000–2021

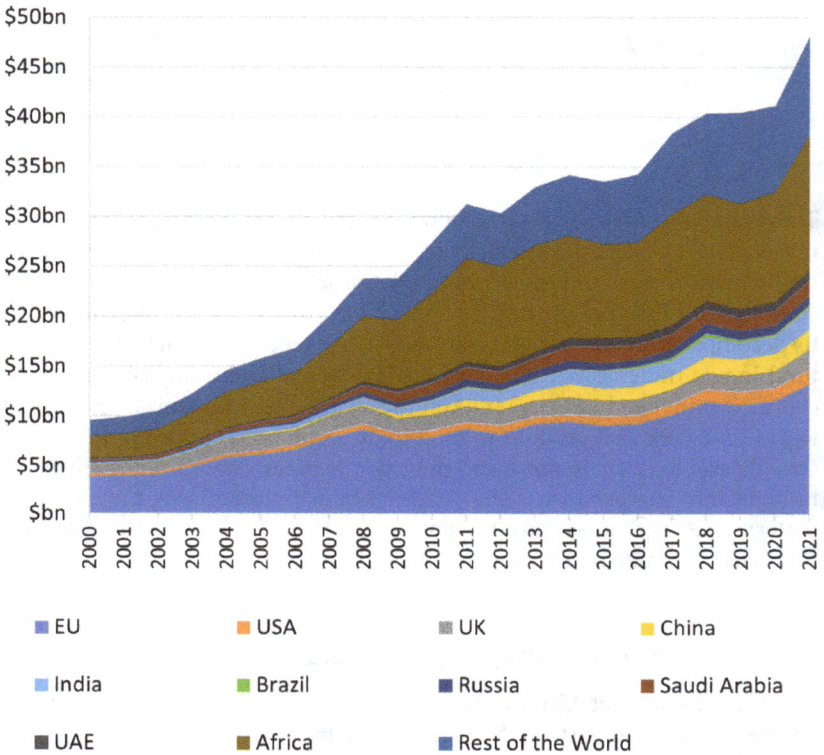

Source: Author's calculations based on UNCTAD (n.d.).

United States (US) have also declined but exports show an upward trajectory. As would be expected, Canada, Ukraine and Russia are important sources of cereal imports. Ukraine supplied 4 per cent of Africa's food imports in 2017–2021, up from 2.1 per cent a decade earlier, while Canada's share has remained constant at 2.4 per cent over the entire period.

Turning to emerging partners, both food exports and imports to and from China and India have increased. In particular, India's share of Africa's food imports has more than doubled – from 2.3 per cent to 4.9 per cent – in the past two decades. Saudi Arabia and the United Arab Emirates (UAE) are growing in significance as export markets. Brazil and to a lesser extent Turkey have emerged as a source of food imports. The rest of the world category, which mainly consists of other developing countries, has also grown in share of Africa's food exports but declined in share of imports. This residual group accounted for an average of 21.1 per cent of the value of Africa's

Figure 8.2: Africa's imports by source country, basic food (US$ billion, current prices), 2000–2021

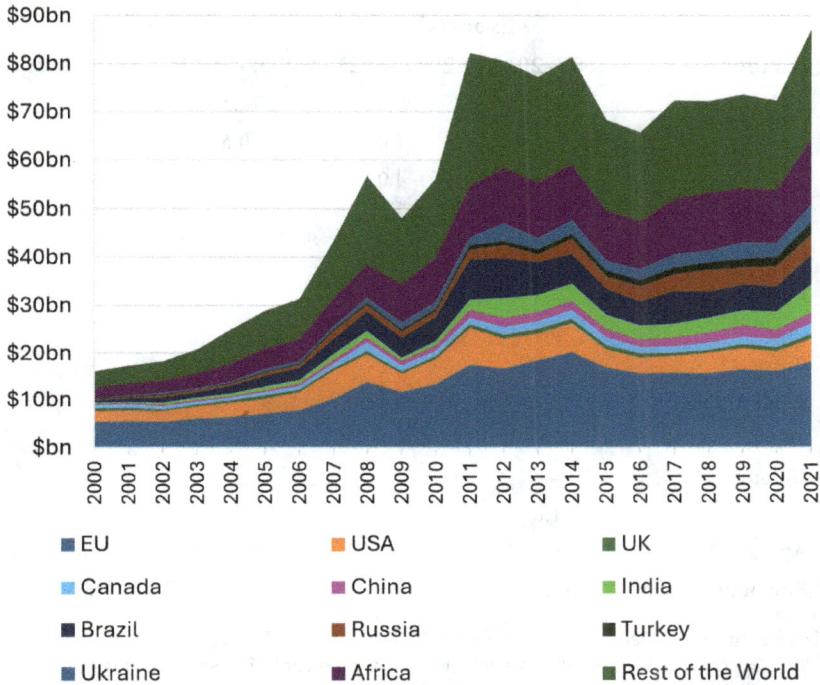

Source: Author's calculations based on UNCTAD (n.d.).

food exports during 2017–2021, up from an average of 16.8 per cent a decade earlier.

Africa's increasing food exports to other developing countries signifies the growing trade ties within the Global South (Table 8.1). The trade relationships often encompass partnerships in agricultural cooperation. For instance, the Gulf states' investments in African farms, which have sometimes been viewed controversially as a 'land grab', bolster local agricultural know-how and production and help to secure local food supplies while also generating exports (Sambidge 2024). (Such land investments do raise important concerns, however, as discussed in Chapter 4.)

Finally, as noted in Chapter 5, intra-African trade in food is significant. Africa absorbed 27.3 per cent of its own food exports during 2017–2021, slightly down from 29.4 per cent a decade earlier. This share is much higher than for total merchandise exports (15 per cent), which suggests that intra-African exports are food-intensive, reflecting the region's comparative advantage and trade complementarity in agriculture.

Table 8.1: Partners' shares of African exports and imports of basic food by value, period averages (%), 2007–2021

Partner	Exports		Imports	
	2007–2011	2017–2021	2007–2011	2017–2021
EU	33.2	27.7	23.3	21.5
USA	2.2	3.0	10.6	6.0
UK	6.5	4.0	1.0	0.9
Canada	—	—	2.4	2.4
China	1.6	4.1	1.8	2.8
India	2.9	4.4	2.3	4.9
Brazil	0.2	0.6	8.6	7.7
Russia	2.3	2.2	2.6	5.3
Saudi Arabia	3.2	3.4	—	—
Turkey	—	—	1.0	2.4
Ukraine	—	—	2.1	4.0
UAE	1.6	2.2	—	—
Africa	29.4	27.3	14.3	15.8
Rest of the World	16.8	21.1	30.1	26.3

Source: Author's calculations based on UNCTAD (n.d.).
Note: '—' means the share is negligible (less than 2 per cent). The shares are calculated on export/import values at current prices.

8.2 Net food trade

Overall, Africa's food import sources are more diversified than its export destinations. In Chapter 2, we noted that in 2021 African countries had an annual net trade deficit of $48 billion in basic foods and of $9 billion in agricultural capital or machinery, while returning a net surplus in exports of agricultural raw materials (including cocoa, tobacco, coffee, tea and spices) of $16 billion and agricultural inputs of $6 billion. We further noted that Africa's imports of basic foods have grown, reaching $104 billion in 2021, up from $97 billion in 2011. Africa's food exports outpaced that growth, increasing gradually but steadily from $41 billion in 2011 to $55 billion in 2021. This yielded the (reduced) net deficit in food trade in 2021, with 42 African states designated as 'net food-importing developing countries' (NFIDCs) by the WTO. How does the deficit breakdown on a bilateral basis?

Table 8.2 ranks Africa's trade partners in descending order of importance as net food providers. The prominence of emerging food trade partners, which include the large rest of the world category alongside the traditional partners, stands out. As expected, all of Africa's major food trade partners are net food suppliers, except the UK.

Table 8.2: Africa's topmost net food suppliers, 2017–2021

Partner	Share of Africa's net food imports (average for 2017–2021, %)
Brazil	16.5
EU	13.8
USA	9.8
Russia	9.1
Ukraine	8.8
India	5.6
Canada	4.3
Turkey	3.7
Africa	1.7
China	1.1
UAE	0.9
UK	−2.9
Rest of the world	27.7

Source: Author's calculations based on data from UNCTAD (n.d.).

Figure 8.3: Share of Africa's net food imports (in percentage) vs. total support to agriculture (US$ billion, current prices), 2017–2021

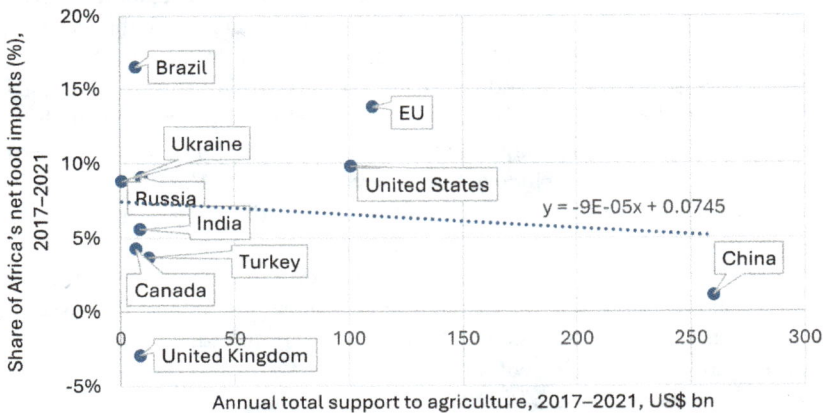

Source: Author's calculations based on OECD (n.d.) and UNCTAD (n.d.).

Although one might expect that countries that subsidise their agricultural sectors would be sources of net food imports for Africa (and the EU and the US, two of the largest subsidisers, are among the largest sources of net imports), there is actually a negative relationship between these countries' total support for agriculture and their net exports to Africa (see Figure 8.3).

8.3 What are the most traded foods?

Table 8.3 summarises the top 10 food products that Africa imports from, and exports to, the EU, the US, the UK, Russia, Brazil, China and India. The selection of countries and products is indicative rather than exhaustive and complements the discussion in Chapter 1. Cereals, dairy, poultry, fish and meat are strongly represented in food imports, with fruit, vegetables, nuts and fish among the leading food exports. Other agricultural exports such as coffee and cocoa are also shown in Table 8.3.

8.4 Trade in agricultural inputs

Inorganic fertilisers and agricultural machinery are not widely used among Africa's smallholder farmers. The former is a boon for sustainability (Baweja, Kumar and Kumar 2020) but the latter can have a negative impact on productivity. Using data from over 22,000 households across Ethiopia, Malawi, Niger, Nigeria, Tanzania, and Uganda, a study found that many as 84 per cent of the farmers surveyed did not use agro-chemical fertilisers and two-thirds of the farmers reported not using inorganic fertilisers. Tractor ownership among the households was minimal (Christiaensen and Demery 2018).

All the same, there is robust trade with traditional and emerging partners in agricultural inputs, including machinery, seeds, fertilisers and herbicides. The benefits are obvious. The availability of improved seeds enhances crop yields, contributing directly to increased food availability and stability. Mechanisation through the use of tractors and modern equipment enhances efficiency, allowing farmers to cultivate larger areas and minimise post-harvest losses. Access to fertilisers and herbicides improves soil fertility, pest management, and crop health, further boosting yields. Experimental analysis of African farms by Duflo, Kremer and Robinson (2008) found that the use of fertilisers and hybrid seed increased maize yields by 40 to 100 per cent. However, low and variable returns on investment in fertilisers continue to limit fertiliser uptake in Africa (Cairns et al. 2021).

Although inorganic fertilisers boost yields, they also cause long-term environmental harm, for example through reductions in plant diversity (CGIAR n.d.; Shi et al. 2024). Organic fertilisers could help to reduce this problem, as they can increase yields without losses of plant diversity (Shi et al. 2024). However, questions remain 'about their long-term impact on soil health and crop productivity' (CGIAR n.d.).

Traditional partners

Besides being the leading food supplier to Africa, the EU is also a major provider of agricultural inputs, notably seeds, agricultural machinery and tractors (Table 8.4). During 2017–2021, Africa sourced over 70 per cent of its agri-

Table 8.3: Africa's top imports and exports of food products (rankings based on average import/export values over 2017–2021)

Imports from:

Brazil	EU	USA	UK	Russia	India	China
Sugar	Wheat	Wheat	Milk and cream	Wheat	Rice	Tea
Maize	Food preparations	Poultry	Food preparations	Barley	Other beet or cane sugar	Rice
Poultry	Milk and cream	Maize, unmilled	Poultry	Fish, frozen	Meat of bovine animals, frozen	Fish, prepared or preserved
Meat of bovine animals, frozen	Poultry	Edible nuts	Vegetables, prepared or preserved	Poultry	Spices	Fish, frozen
Rice	Fish, frozen	Oil cake and other solid residues		Yeasts	Bread, pastry, cakes etc.	Vegetables, prepared or preserved
Sausages	Malt	Leguminous vegetables		Leguminous vegetables	Sugar confectionery	Food preparations
Pig meat	Cheese, curd	Milk and cream		Other maize, unmilled	Extracts, essences and concentrates of coffee	Sausages and preparations thereof
	Bread, pastry, cakes etc.	Edible animal offal		Meat of bovine animals, frozen	Sugars, of beet or cane	Other fresh or chilled vegetables
	Bovine animals, live	Grain sorghum, unmilled		Oil cake and other solid residues	Coffee, not roasted	Yeasts
				Other sugars	Food preparations	Sugar confectionery

(Continued)

Table 8.3: (Continued)

Exports to:

Brazil	EU	US	UK	Russia	India	China
Cocoa beans	Cocoa beans	Cocoa beans	Fruit, fresh or dried	Oranges and other citrus fruit	Edible nuts	Oranges and other citrus fruit
Fish, frozen	Fruit, fresh or dried	Cocoa paste	Fresh or chilled vegetables	Cocoa beans	Leguminous vegetables	Edible nuts
Vegetables, prepared or preserved	Coffee	Coffee	Grapes	Fruit, fresh or dried	Spices	Residues of starch
Cocoa paste	Fish, prepared or preserved	Spices	Oranges	Cocoa paste	Oranges and other citrus fruit	Coffee, not roasted
Cocoa powder	Oranges and other citrus fruit	Edible nuts	Fish, prepared or preserved	Potatoes, fresh or chilled	Coffee, not roasted	Fruit and nuts
Spices	Cocoa paste	Oranges and other citrus fruit	Cocoa beans	Tea	Oil cake and other solid residues	Meat of bovine animals, frozen
Fruit, fresh or dried	Other vegetables	Fish, prepared or preserved	Tea	Other citrus fruit, fresh or dried	Cocoa beans	Leguminous vegetables
Oranges and other citrus fruit	Cocoa butter	Vegetables, prepared or preserved	Tomatoes	Grapes, fresh or dried	Fruit, fresh or dried	Cocoa beans
Fruit and nuts	Tomatoes	Fruit, fresh or dried	Apples	Coffee, not roasted	Tea	
	Grapes	Sugar	Cocoa butter	Other fresh or chilled vegetables		

Source: Author's calculations based on UNCTAD (n.d).

Table 8.4: Top five non-African exporters of agricultural inputs and machinery to Africa, average values for 2017–2021 (US$ million)

Country	Agricultural machinery and parts (Standard International Trade Classification) (SITC) 721)	Country	Fertilisers (SITC 562)	Country	Seeds, fruits and spores for planting (SITC 2925)	Country	Herbicides (SITC 5913)	Country	Tractors (SITC 7125)
EU	527.6	EU	429.6	EU	306.3	China	691.0	EU	191.8
China	141.0	Russia	304.0	USA	35.1	EU	170.4	USA	137.0
USA	102.0	Saudi Arabia	242.0	Israel	12.0	India	62.9	India	133.0
India	62.1	China	219.0	Australia	9.8	USA	31.9	China	70.8
Turkey	56.0	Jordan	122.0	India	6.8	Israel	8.4	UK	45.9
Rest of the world	216.2	Rest of the world	1762.0.0	Rest of the world	66.2	Rest of the world	92.8	Rest of the world	154.3
Total imports	**1104.9**	**Total imports**	**3042.5**	**Total imports**	**436.4**	**Total imports**	**1011.3**	**Total imports**	**732.8**

Source: Author's calculations based on data from UN Comtrade.
Note: Numbers for total imports may not match the totals for those from individual partners because the partner jurisdiction is not identified for some imports, and because of rounding off.

cultural seeds from the EU, underpinned by collaborative arrangements such as the Alliance for a Green Revolution in Africa (AGRA) and the Seed and Knowledge Initiative (AGRA 2021). Other initiatives, such as the EU–Africa Partnership on Food and Nutrition Security and Sustainable Agriculture, also support African farmers' access to inputs, among other objectives (Partnership on Food and Nutrition Security and Sustainable Agriculture (FNSSA) n.d.). In 2017–2021, about half of Africa's imports of agricultural machinery and a third of its tractor imports came from the EU.

Among other traditional partners, the US is a notable supplier of tractors and agricultural machinery to Africa, accounting for about one-fifth of Africa's tractor imports in 2017–2021 according to UN Comtrade data. The US Feed the Future programme seeks to improve agricultural production and markets in developing countries (USAID n.d.).

The UK is a significant supplier of tractors and an important exporter of agricultural machinery and parts to Africa. The UK's involvement in African agriculture can be traced back to the colonial era, when British companies established agricultural plantations and introduced modern farming techniques. This influence has a left a lasting legacy, as British agricultural expertise and machinery continue to be used in many parts of the continent.

Emerging partners

China has emerged as the largest supplier of herbicides to the continent and is next to the EU in agricultural machinery exports to Africa. Indian tractors are becoming popular on African farms, not only because of price and durability but also because of their adaptability to local agricultural conditions. Saudi Arabia (fertilisers), Israel (seeds) and Turkey (tractors) are important input suppliers to Africa.

Much like the traditional partners, Africa's emerging partners have set up initiatives to promote access to agricultural inputs and technology, knowledge-sharing and capacity-building (*Business Times* 2018). Notable among these are the India's Technical and Economic Cooperation Programme, the China–Africa Agriculture Cooperation Programme, and collaboration between the Brazilian Agricultural Research Corporation (Embrapa) and some African countries (Santos 2016).

8.5 What are the trade policy regimes underpinning Africa's bilateral food trade?

This section considers the trade policies and related issues that impact bilateral food trade with selected traditional and emerging trading partners. Among traditional partners, the EU and US have long established trade policy regimes that impact Africa's food trade. Russia and Ukraine also have a long

Box 8.1: The Black Sea Grain Initiative (BSGI)

The BSGI was negotiated in July 2022 between Turkey, Russia, Ukraine and the UN as a means of ensuring that Ukraine could ship its grain via the Bosphorus. The deal ended one year later, in July 2023, as Russia retaliated against Western sanctions and attacks by Ukraine on its Black Sea fleet.

Under the initiative, Ukraine exported over 33 million tonnes of grain between July 2022 and July 2023. Partly as a result of this, the price of grain stabilised at $800 per tonne, down from a high of $1,360. With the collapse of the initiative, Russia announced it would donate 'free grain' to six countries with which it has strong links: Somalia, Burkina Faso, Eritrea, Zimbabwe, Central African Republic and Mali. This prompted UN warnings that a 'handful of donations' would not correct the 'dramatic impact' caused by the end of the Black Sea deal. The suspension of the BSGI again generated fluctuations in international wheat prices. Djibouti, Somalia and Sudan – highly dependent on imports through the Black Sea – were particularly vulnerable. Somalia's reliance on Ukraine for over 60 per cent of its wheat imports underscores this vulnerability and resulted in an urgent search for alternative sources for wheat supply, including through humanitarian aid. In Sudan, a decline in wheat production in 2023 amid political instability exacerbated the effects of the collapse of the BSGI. A dramatic rise in local wheat flour prices followed, reducing access and affordability. The uncertainties following the Russia–Ukraine conflict led most African countries that depended on grain supplies from the warring parties to diversify their sources of imports.

Source: WFP (World Food Programme) (2023); Wintour (2023).

history of trade with African countries, supplying cereals. Consequently, the Russia–Ukraine war, including the rise and demise of the Black Sea Grain Initiative (see Box 8.1), exposed vulnerabilities for some African countries. Among emerging trading partners, China, India and Brazil illustrate the growing trade partnerships with the Global South.

Traditional partners – the EU

The EU's bilateral trade arrangements with African countries vary according to geographical location on the African continent and level of development (Luke, McCartan-Demie and Guepie. 2023). Specifically, as regards food trade, agricultural protectionism, food safety standards, intellectual property rights, and initiatives emanating from the EU's Green Deal are problematic areas in the bilateral food trade relationship.

EU agricultural protectionism is exercised through its agriculture tariff schedules, domestic support or subsidies for its farmers and recourse to 'special agricultural safeguards', all of which are permitted under WTO rules. Since the start of the Doha Round in 2001, export subsidies have been virtually eliminated and import tariffs on agricultural products have been significantly reduced. However, the share of domestic subsidies in total support to farmers in OECD countries has more than doubled in the two decades since then (Anderson et al. 2021, p.1). As discussed in Chapter 9, rich countries' agricultural subsidies incentivise production, which contributes to global food availability but disincentivises production in poorer and net food-importing countries. This presents a major challenge to African agricultural production, trade and food security, as do the persistent imbalances in the WTOs Agreement on Agriculture (Eagleton-Pierce 2012; ECA Southern Africa Office (SRO-SA) 2007; Singh 2017, cited in Hopewell 2022).

In relation to tariff schedules, in the dairy sector, average tariffs are as high as 32 per cent, with sugar and confectionery at 27 per cent, meat at 19 per cent, cereals and cereal preparations at 17 per cent, and fruits and vegetables at 13 per cent (WTO 2019). EU agricultural tariff-rate quotas (TRQs)[1] are quite diverse and apply to a wide range of agricultural goods such as meat, dairy, cereals, fruit and vegetables, and processed foods, some of which are of major export interest to Africa. However, the utilisation rate of these TRQs has remained low and constant, averaging 39 per cent in recent years (WTO 2023). The evidence suggests that TRQs can facilitate market access for African horticultural exports to the EU, such as South Africa's exports of canned fruit, but they pose challenges due to their limited nature and potential for market distortions (Muchopa 2021).

As for domestic support, annual spending on EU farm subsidies is a multiple of the gross domestic product of many African countries. In 2019–2020 this was €81 billion and applied to farmers' income support, rural development and market measures (Directorate-General for Agriculture and Rural Development n.d. a; Directorate-General for Agriculture and Rural Development n.d. b). Farm subsidies incentivise overproduction and contribute to higher greenhouse gas (GHG) emissions in sectors such as meat production.

Special agricultural safeguards[2] are applied to three groups of products: sugar, fruits and vegetables, and poultry and meat, which are already aided by subsidies and tariff protection, including specific seasonal tariffs (WTO 2019). This is why Everything but Arms, the EU's concessional trade arrangement for least-developed countries (LDCs), is often mocked as Everything but Farms! The combined effect of these policies is that the EU, which might well be a net importer of some of these products, is actually a net exporter. An example is Morocco's food trade with the EU. Morocco is more competitive than EU producers in certain fruits and vegetables such as tomatoes, oranges and clementines. However, EU subsidies distort farm prices, making it difficult for Morocco and other North African producers to compete in the EU market (van Berkum 2013).

The EU's food safety standards are modelled on the WTO Sanitary and Phytosanitary Measures (SPS) Agreement and encompass various aspects of the food value chain, including hygiene, labelling, pesticide residues, contaminants and traceability. Based on its Farm to Fork Strategy, the EU's SPS regime is recognised as going beyond the protection of consumer health. The strategy has a wide compass that includes animal welfare, sustainable agricultural practices, environmental protection and nature conservation (Directorate-General for Health and Food Safety 2024). An empirical study by UNCTAD found that the EU's SPS measures resulted in higher burdens for lower-income countries and a 14 per cent reduction in their agricultural exports (Murina and Nicita 2014). In 2022, South Africa filed a complaint under the WTO's dispute settlement arrangements against the EU on what it considered to be unwarranted phytosanitary requirements for its fruit exports (van der Ven and Luke 2023).

Intellectual property rights as they relate to plant genetic resources and technology transfer is another area of concern in Africa's bilateral food trade relationship with the EU. As previously noted, the EU is a source of key agricultural inputs, including seeds. EU intellectual property rights requirements restrict farmers' ability to save and exchange seeds. In addition, 'non-complying seeds, including traditional heterogeneous varieties, are banned' (de Mévius 2022). Notwithstanding certain derogations (such as the right of farmers to reuse and multiply patent-protected seeds for use on their own farms), the legal space for the conservation and sustainable use of plant genetic resources for food agriculture is narrow (de Mévius 2022; Gil-Robles and Edlinger, 1998). Although this legislation applies only to EU member states, in some cases it has inspired other countries to adopt similar legislation;[3] Egypt, Morocco and Tunisia have done this as part of signing the Euro-Med Association Agreements with the EU (Peschard, Golay and Araya 2023, p.45). Moreover, the EU Seed Marketing Legislation that is being developed prohibits public gene banks, private collections, and unauthorised use of EU-originating seeds (de Mévius 2022).[4]

During the early years of the 2020s, the EU elaborated a range of policy initiatives under its Green Deal and Fit for 55 climate package, aimed at reducing carbon emissions by 55 per cent by 2030 and achieving carbon neutrality by 2050. Among these are the Carbon Border Adjustment Mechanism (CBAM) and the Deforestation Regulation. Of these two policy measures, the proposal for a CBAM only marginally affects agriculture and food trade, although further measures on agricultural products have not been ruled out in the future. In the first phase of the scheme, which came into effect in October 2023, the CBAM introduced a levy on emissions embedded in imported goods such as cement, aluminium, iron and steel, fertilisers, electricity and hydrogen to address the issue of 'carbon leakage'. This occurs where EU-based producers are subjected to its emissions trading scheme while imports may not face the same level of levies on emissions. Of the products included in the scheme that are directly relevant to agriculture, only the small trade in fertilisers exported to the EU from countries such as Mauritania and Morocco is initially affected.

However, the EU Deforestation Regulation, which aims to address the environmental impacts of deforestation and forest degradation associated with EU imports and production of specific agricultural commodities (Regulation on Deforestation-Free Products n.d.), will have a direct impact on bilateral food and agricultural trade. The regulation – which was initially scheduled to come into effect on 30 December 2024 but delayed for a year to 30 December 2025 – targets products with high deforestation risk such as cocoa, coffee, palm oil, soya, beef, wood and rubber. The scope could be extended to include pig meat, sheep, goats, poultry, maize, charcoal and printed paper products. Importers of the covered goods into the EU must ensure that these products do not come from land that was deforested after 31 December 2021, produced in accordance with both the laws of the country of origin and international law, and respect the rights of traditional communities over their territories. To facilitate compliance, the EU has created a benchmarking system categorising countries into low, standard or high risk of deforestation. Low-risk countries will have simplified due diligence obligations, reducing compliance costs for EU importers. High-risk commodity-exporting countries will face more rigorous scrutiny. Establishing risk and the traceability of products including through the satellite and GPS technologies that are essential to the scheme will impose additional costs on African food and agricultural exporters.

However, these climate-focused interventions may also have benefits for Africa by impacting the pace of climate change, given how far the continent is expected to suffer (and is already suffering) as a result of the climate crisis (World Meteorological Organization 2020; World Meteorological Organization 2023).

Traditional partners – the US

The Africa Growth and Opportunity Act (AGOA), which has been in effect since 2001, providing eligible African countries south of the Sahara with duty-free access to the US market for over 6,700 products, is the main trade policy framework for bilateral trade with the US. Good governance is a major criterion for eligibility. At the time of writing, six African countries (Burkina Faso, Guinea, Mali, Niger, Sudan and Ethiopia) have been suspended from the scheme for not being compliant with the governance criterion. The risk of suspension of AGOA benefits generates uncertainty for investors and exporters. For non-AGOA eligible African countries, trade with the US is carried out under most-favoured nation tariffs or the US Generalised System of Preferences (GSP). Since 2006, Morocco has had a free trade area arrangement with the US. In relation to food, wheat is a major import to Morocco from the US, with fruit, nuts and horticulture produce going in the other direction (Office of the United States Trade Representative n.d.).

AGOA has been reauthorised by the US Congress five times: in 2004, 2006, 2007, 2012 and 2015. While the earlier extensions were short-term, the 2015 extension was for 10 years, allowing greater predictability in trade and investment decisions. At the time of writing, discussions have begun for a further extension in 2025. Its extension is shrouded in uncertainty given the second Trump administration's aggressive transactional approach to trade policy.

As we saw earlier in this chapter in the discussion of bilateral trade flows, AGOA has facilitated a modest growth in Africa's food and agricultural exports. Africa's agricultural exports to the US have increased by 60.8 per cent in the past 10 years to reach US$2.9 billion in 2022. However, agricultural products account for just 11 per cent of non-oil imports under AGOA (Schneidman, McNulty and Dicharry 2021). Other challenges to food and agricultural exports under AGOA include product exclusion and erosion of preferences as market access concessions are granted by the US to an increasing range of countries, Viet Nam for example. Capacity to comply with non-tariff barriers, particularly SPS regulations, has also hindered AGOA agricultural exports. For example, lengthy US import approval procedures for horticultural products meant that baby squash and courgettes from Zambia, which were considered for export following the enactment of AGOA in 2001, received the green light more than seven years later in December 2008 (Pasco 2010).

As in the EU, agricultural protectionism is exercised through high import tariffs for farm products and subsidies for farmers. This makes some African exports less competitive in the US market. High tariffs and TRQs permeate several agricultural sectors that also attract substantial farm subsidies, including sugar, tobacco, cotton, dairy and beef. The US maintains 46 TRQs on seven commodities (Meltzer 2015).

Peanuts, for example, attract over-quota tariffs of up to 163.8 per cent. This is a prohibitive tariff that shuts out any prospect of African peanut exports to the US beyond the quota amount, since imports beyond the quota do not benefit from duty-free access under AGOA. Tobacco faces an ad valorem tariff[5] equivalent of 350 per cent, which is a high barrier to overcome for Malawi's tobacco to enter the lucrative US market. Dairy products attract the highest number of TRQs (22) across 107 in-quota tariff lines, with ad valorem equivalents ranging from 30 to 120 per cent. Sugar, a major African export to the US as noted from Table 8.3, is hit with over-quota tariffs of up to 210 per cent. One study estimates that the complete elimination of US tariffs on agricultural exports under AGOA would increase African exports by more than $105 million while reducing US production by less than US$10 million (Mevel, Lewis and Kamau 2013).

In 2019–2020, the United States provided €190.6 billion in farm subsidies (Directorate-General for Agriculture and Rural Development n.d. b). These covered more than 150 programmes including ad hoc disaster assistance, agricultural risk, crop insurance, conservation, price loss below the products' reference price, marketing and export aid, and research and development (Edwards 2023). Here again, these subsidies are allowed under WTO rules, as discussed in Chapter 9. The scale of the subsidies encourages overproduction and generates higher carbon emissions in some sectors. The subsidies also have the effect of out-competing African food exports in sectors where they are competitive, notably beef, maize, soya beans and dairy. A convenient outlet for overproduction is food aid. Quite apart from humanitarian and emergency relief, food aid as discussed in Chapter 4 and Chapter 9 can undermine local production and generate dependencies.

Traditional partners – Russia and Ukraine

Both Russia and Ukraine are major players in global agricultural production and trade in cereals. Ukraine is also a major producer of sunflower oil. Russia is Africa's third biggest supplier of fertilisers. Cereals represented 35 per cent of Africa's imports from Russia during 2017–2021. According to UNCTAD-STAT data, Ukraine's share of Africa's food imports doubled to 4 per cent in the past decade. Cereals from Ukraine represent 10 per cent of Africa's world cereal imports. Africa's exports to Ukraine are negligible but about half of Africa's exports to Russia during 2017–2021 were in food products, notably agricultural commodities, fruit and horticulture.

Africa maintains a deficit in net food trade with both Russia and Ukraine. However, trade with Russia and Ukraine is concentrated in a handful of African countries, namely Egypt, Kenya, Sudan, Tunisia, Ethiopia, Somalia and Djibouti. This explains the limited effect on Africa as a whole of the supply disruption that followed Russia's invasion of Ukraine in 2022. Although food imports from Russia and Ukraine are small in relation to Africa's total food imports, their concentration in a few countries resulted in apprehensions about the availability of supplies, resulting in the negotiation of the Black Sea Grain Initiative (see Box 8.1).

Unlike the EU and the US, Russia and Ukraine do not have well-defined trade policy frameworks with African countries. As a member of the Eurasian Customs Union (with Belarus, Kazakhstan, Kyrgyzstan and Amenia), Russia offers preferential market access to developing countries through a GSP scheme and participates in the WTO's duty-free quota-free (DFQF) market access for least-developed countries.

Russia also engages in technical assistance and knowledge-sharing activities, including technology transfer, research collaboration and agribusiness development. A forum on agribusiness was held during the 2019 Russia–Africa summit (Yakovenko 2019).

Emerging partners – China

As previously noted, China–Africa trade in agricultural goods is modest but growing. During 2017–2021, China accounted for 4.1 per cent of Africa's food exports and 2.8 pr cent of imports. In comparison with trade in all goods, China accounted for 15.2 per cent of Africa's exports and 17.3 per cent of imports, making agricultural trade a small part of total trade. Fruits, nuts, vegetables and beef are among the food exports to China, along with coffee, tobacco and cotton among other agricultural products. Imports include rice, food preparations, yeasts, sugar, agricultural inputs and machinery. Chinese investment has been made in trade-related infrastructure, such as transportation and storage facilities (Hamilton and Maliphol 2021). Technical assistance, technology transfer, knowledge-sharing and capacity-building initiatives are directed to African farmers and agribusinesses (Ministry of Foreign Affairs of the People's Republic of China 2024).

In relation to trade policy, only the 'most basic' framework exists for trade between China and African countries (Luke, McCartan-Demie and Guepie 2023). As it considers itself a developing country (and is still an upper-middle-income country according to the World Bank's classification, rather than a high-income one), China does not offer a GSP (World Bank n.d.). However, since 2010 it has participated in the WTO's duty-free quota-free scheme for LDCs for up to 98 per cent of tariff lines. In 2021, China concluded a free trade agreement (FTA) with Mauritius, the only trade deal it has with an African country. The make-up of the Mauritian economy and its highly liberalised trade regime, and the challenges for Mauritian firms to increase their exports to China in spite of the FTA, are such that the agreement will have little impact on food and agricultural trade (Ancharaz and Nathoo 2022).

Like other large economies, China protects its agricultural sector through the use of such tools as tariffs, subsidies and food safety measures. Concerns that the latter have been a major impediment to Africa's exports of agricultural products have prompted African countries to negotiate 'green lanes' with China to ease the process of carrying out phytosanitary assessments in exporting agricultural produce to China.

Emerging partners – India

India–Africa trade has a long history, facilitated by the shared geography of the Indian Ocean rim. Engagement on trade is also driven by the presence of a large Indian diaspora on the continent (Ben Barka 2011; Chakrabarty 2016).

In line with growing trade ties between Africa and countries in the Global South, food trade with India has grown during the last two decades. India accounted for 4.4 per cent of Africa's food exports and 4.9 per cent of food imports during 2017–2021. The composition of exports to India are comparable with those to China and consist of nuts, fruits, spices, vegetables and agricultural commodities like coffee and cocoa. Imports from India include rice, sugar, meat and food preparations.

Agriculture is a strategic sector in India and protected by policy measures allowed by the WTO. As a developing country, India like China does not have a GSP scheme but participates in the WTO DFQF initiative for LDCs. This allows duty-free treatment for up to 98 per cent of tariff lines as of 2014 (Ancharaz and

Ghisu 2014). But some agricultural commodities in which African countries are competitive, such as coffee and tea, are excluded from the scheme (Ancharaz and Ghisu 2014; Ancharaz, Ghisu and Wan 2014, p.25). Research in Ethiopia, Tanzania and Uganda has found that uptake of the scheme was marred by lack of awareness of the scheme among exporters of the opportunities it offered (Ancharaz, Ghisu and Frank 2014a, p.11; Ancharaz, Ghisu and Frank 2014b, p.26; Ancharaz, Ghisu and Wan 2014, p.25).

Emerging partners – Brazil

Brazil's emergence as an agricultural superpower is evidenced by its leading role as a global supplier of soya, meat, grains and sugar. In Africa, Brazil is the biggest supplier of sugar, maize and poultry, and among the top exporters of other animal products. During 2017–2021, Brazil accounted for 7.7 per cent of Africa's food imports, making it more important than all other trading partners except the EU (as a bloc) and the African continent itself (as a whole). Food represents two-thirds of all imports from Brazil, while Africa's food exports are negligible.

As a developing country, Brazil does not offer trade preferences to African countries. In 2016, Mercosur, of which Brazil is a member, and the Southern Africa Customs Union (SACU), which includes South Africa, Botswana, Namibia, Eswatini and Lesotho, concluded an FTA. It is a shallow agreement that sets out preference margins of 10, 25, 50 and 100 per cent on 1,050 tariff lines covering both industrial and agricultural goods (Ministério das Relações Exteriores [Ministry of Foreign Affairs], 2016).

Brazil shares its agricultural know-how through robust technical assistance outreach. Similar agronomic conditions and affinities between Africa and Brazil have often been invoked to support the transfer of knowledge and technology between the two partners. This includes initiatives such as the Brazil–Africa Agriculture and Food Security Programme, which seeks to foster self-reliance in African agriculture by promoting sustainable practices and agribusiness development (World Food Programme 2020), and More Food International (MFI), a cooperation programme aimed at strengthening the productive capacity of African smallholder farmers. However, a case study of the adoption of MFI in three Africa countries – Ghana, Mozambique and Zimbabwe – suggests that the programme has not worked well as local conditions were not taken fully into account (Cabral et al. 2016).

Summary

This chapter has reviewed the food trade relationships between Africa and its traditional bilateral partners such as the EU, the US, Russia and Ukraine and emerging bilateral partners such as Brazil, China and India. The chapter has uncovered various aspects of how these interactions impact food trade and food security on the African continent in terms of both the value of net imports and the specific products that are provided. The geography of these relationships is changing, with increasing food trade flows between African and emerging partners in the Global South. Traditional partners are also losing trade shares in agricultural inputs such as machinery, seeds, fertilisers and herbicides to emerging partners.

Brazil is the largest net food supplier to Africa, followed by the EU and the US. The EU, however, remains Africa's most important market for both food exports and imports. The EU and the US are also significant suppliers of agricultural seeds, machinery and tractors to Africa. Among traditional partners, Russia and Ukraine are a major source of cereal exports to Africa. The Russia–Ukraine war that started in 2022 disrupted the flow of these exports. But the concentration of Russia's and Ukraine's grain exports in a few African countries limited a wider damaging effect, although the collapse of the BSGI after only one year in 2023 resulted in a surge in wheat prices.

In assessing the trade policy regimes that underpin trade flows, we saw that many African countries benefit from market access concessions such as the EU's Everything but Arms, the US's AGOA and the WTO's DFQF initiative. But there is a high level of agriculture sector protectionism in bilateral partners' markets through measures allowed by WTO rules. These include high import tariffs for farm products and subsidies to farmers, which lead both to overproduction and to enhanced levels of GHG emissions in some food production sectors. Agricultural protectionism makes many African food exports less competitive, especially in traditional partners' markets. Capacity in several African countries to meet food safety standards is a perennial challenge. In the case of China, significant efforts have been made to work with African exporters to ease this difficulty through the introduction of 'green lanes'. Policies in the EU related to its Green Deal and Fit for 55 such as the CBAM and Deforestation Regulation will increasingly expose the nexus between trade and climate to greater scrutiny.

Forty-two of the 54 African countries are net food importers, which elevates bilateral food trade to a matter of strategic importance for these countries. Most of these countries are part of the NFIDC group at the WTO. These countries coordinate efforts to keep international food markets open, monitor food aid flows and constitute an important stakeholder group in negotiations to reform WTO rules on agriculture.

Notes

[1] In the EU, 'tariff-rate quotas' refers to quotas for imports than can benefit from a lower tariff than any imports that exceed the quota (European Commission 2024).

[2] 'Special agricultural safeguards' refers to temporary restrictions on imports used to deal with special circumstances such as a sudden surge in imports (World Trade Organization 2004).

[3] As of 2022, Egypt, Ghana, Kenya, Morocco, Tanzania and Tunisia were party to the 1991 UPOV Convention. Farmers in a state party to this convention 'cannot save or reuse seeds of protected varieties, except on their own farms, and only provided that their government has adopted an optional exception to this effect (Articles 15). Moreover, this exception must be "within reasonable limits" and safeguard "the legitimate interests of the breeder." This means, for example, that it can be limited to certain crops or can be conditional on the payment of license fees' (Peschard, Golay and Araya 2023, p.21, based on UPOV International Union for the Protection of New Varieties of Plants 2009, pp.8–11).

[4] See also ARC (2023).

[5] An ad valorem tariff is one where the tariff to be paid is determined as a percentage of the value of the goods being imported.

References

AGRA (2021) Alliance for a Green Revolution in Africa. https://agra.org

Ancharaz, Vinaye; and Ghisu, Paolo (2014) 'Deepening India's Engagement with the Least Developed Countries: An In-Depth Analysis of India's Duty-Free Tariff Preference Scheme', International Centre for Trade and Sustainable Development.

Ancharaz, Vinaye; Ghisu, Paolo; and Frank, Nicholas (2014a) 'Ethiopia: Deepening Engagement with India through Better Market Access', International Centre for Trade and Sustainable Development. https://perma.cc/JN4Z-JDS9

Ancharaz, Vinaye; Ghisu, Paolo; and Frank, Nicholas (2014b) 'Tanzania: Deepening Engagement with India through Better Market Access', Geneva: International Centre for Trade and Sustainable Development. https://perma.cc/82UE-K4E2

Ancharaz, Vinaye Dey; Ghisu, Paolo; and Wan, Jessica (2014) 'Uganda: Deepening Engagement with India through Better Market Access', International Centre for Trade and Sustainable Development [Preprint]. https://doi.org/10.7215/co_ip_20141104b

Ancharaz, Vinaye; and Nathoo, Rajiv (2022) 'Mauritius' Free Trade Agreement with China: Lessons and Implications for Africa', Policy Insights, vol. 127. https://www.jstor.org/stable/resrep41902

ARC (2023) 'EU Seed Law Reform and New Genetic Engineering – Double Attack on our Seeds', Agricultural and Rural Convention. https://perma.cc/E4ND-CQZ4

Baweja, Pooja; Kumar, Savindra; and Kumar, Gaurav (2020) 'Fertilizers and Pesticides: Their Impact on Soil Health and Environment', in Giri, B. and Varma, A. (eds) *Soil Health*, Cham: Springer, pp.265–85. https://doi.org/10.1007/978-3-030-44364-1_15

Ben Barka, Habiba (2011) 'India's Economic Engagement with Africa', *Africa Economic Brief*, 2(6). https://perma.cc/P9PH-Q8VS

Business Times (2018) 'Zim Stocks Up Mahindra Tractors as Farm Mechanisation Gains Momentum', 23 August. https://perma.cc/M9KS-SD4A

Cabral, Lídia; Favareto, Arilson; Mukwereza, Langton; and Amanor, Kojo (2016) 'Brazil's Agricultural Politics in Africa: More Food International and the Disputed Meanings of "Family Farming"', *World Development*, vol. 81, pp.47–60. https://doi.org/10.1016/j.worlddev.2015.11.010

Cairns, Jill E.; Chamberlin, Jordan; Rutsaert, Pieter; Voss, Rachel C.; Ndhlela, Thokozile; and Magorokosho, Cosmos (2021) 'Challenges for Sustainable Maize Production of Smallholder Farmers in Sub-Saharan Africa', *Journal of Cereal Science*, vol. 101, p.103274. https://doi.org/10.1016/j.jcs.2021.103274

CGIAR (n.d.) 'Balancing food security with environmental sustainability: CGIAR approach to nutrient management, food security, and environmental sustainability https://perma.cc/BC5M-EB3K

Chakrabarty, Malancha (2016) 'Understanding India's Engagement with Africa', *Indian Foreign Affairs Journal*, 11(3), pp.267–280. http://www.jstor.org/stable/45341961

Christiaensen, Luc; and Demery, Lionel (2018) 'Agriculture in Africa: Telling Myths from Facts', World Bank. http://hdl.handle.net/10986/28543

de Mévius, Joséphine (2022) 'Impact of the European Union's Seed Legislation and Intellectual Property Rights on Crop Diversity', *European Energy and Environmental Law Review*, vol. 31, no. 3, pp.149–62. https://doi.org/10.54648/eelr2022010

Directorate-General for Agriculture and Rural Development (n.d. a) 'The Common Agricultural Policy at a Glance', European Commission | Agriculture and Rural Development. https://perma.cc/7M23-5RJV

Directorate-General for Agriculture and Rural Development (n.d. b) 'The WTO and EU Agriculture', European Commission | Agriculture and Rural Development. https://perma.cc/GPD8-VUQA

Directorate-General for Health and Food Safety (2024) 'Food Safety', European Commission | Food, Farming, Fisheries. https://perma.cc/UUZ2-WTHV

Duflo, Esther; Kremer, Michael; and Robinson, Jonathan (2008) 'How High
   Are Rates of Return to Fertilizer? Evidence from Field Experiments in
   Kenya', *American Economic Review*, vol. 98, no. 2, pp.482–88.
   https://doi.org/10.1257/aer.98.2.482

Eagleton-Pierce, Matthew (2012) '5 The heretics in the house',
   in M. Eagleton-Pierce (ed.) *Symbolic Power in the World Trade
   Organization*. Oxford: Oxford University Press.
   https://doi.org/10.1093/acprof:oso/9780199662647.003.0005

ECA Southern Africa Office (SRO-SA) (2007) *Impact of Food Aid
   and Developed Countries' Agricultural Subsidies on Long-Term
   Sustainability of Food Security in Southern Africa*. ECA/SA/FOO-
   DAID/2007/1. Addis Ababa: Economic Commision for Africa.
   https://repository.uneca.org/bitstream/handle/10855/14942/Bib.%20
   56979_I.pdf?sequence=1&isAllowed=y

Edwards, Chris (2023) 'Cutting Federal Farm Subsidies', CATO Briefing
   Paper no. 162. https://perma.cc/H8MS-MYGT

European Commission (2024) 'Tariff Rate Quotas', European Commission,
   Agriculture and Rural development. https://perma.cc/4P2M-5JLM

Gil-Robles, J. M. and Edlinger, R. (1998) *Directive 98/44/EC of the European
   Parliament and of The CounciL of 6 July 1998 on the Legal Protection of
   Biotechnological Inventions.*
   https://eur-lex.europa.eu/legal-content/EN/TXT/PDF/?uri=CELEX
   :31998L0044&from=EN

Hopewell, Kristen (2022) 'Heroes of the developing world?
   Emerging powers in WTO agriculture negotiations and dispute
   settlement', *The Journal of Peasant Studies*, 49(3), pp.561–584.
   https://doi.org/10.1080/03066150.2021.1873292

Hamilton, Clovia and Maliphol, Sira (2023) 'Reimagining China's
   Transportation Funding Investments in Africa in the Context of
   COVID-19', *Transportation Research Record*, 2677(4), pp.118–128.
   https://doi.org/10.1177/03611981211031228

Luke, David; McCartan-Demie, Kulani and Guepie, Geoffroy (2023).
   "Africa's trade arrangements with the European Union and China,"
   in D. Luke (ed), *How Africa Trades*, London: LSE Press, p.51–76.
   https://doi.org/10.31389/lsepress.hat

Meltzer, Joshua P. (2015) 'Reforming the African Growth and Opportunity Act
   to Grow Agriculture Trade', Brookings. https://perma.cc/DY82-YBGV

Mevel, Simon; Lewis, Zenia; and Kamau, Anne (2013) 'African the
   Growth and Opportunity Act an Empirical Analysis of the Possibilities
   Post-2015', Brookings. https://perma.cc/4DF8-VZJA

Ministério das Relações Exteriores (2016) 'Entry into force of the Mercosur-SACU Preferential Trade Agreement', gov.br. https://perma.cc/932H-ZFBE

Ministry of Foreign Affairs of the People's Republic of China (2024) 'China-Africa cooperation empowers African countries' drive toward modernization', The 2024 Summit of the Forum on China-Africa Cooperation. https://perma.cc/764T-UNE6

Muchopa, Chiedza L. (2021) 'Economic Impact of Tariff Rate Quotas and Underfilling: The Case of Canned Fruit Exports from South Africa to the EU', *Economies*, vol. 9, no. 4, p.155. https://doi.org/10.3390/economies9040155

Murina, Marina and Nicita, Alessandro (2014) 'Trading with Conditions: the Effect of Sanitary and Phytosanitary Measures on Lower Income Countries' Agricultural Exports', *Policy Issues in International Trade and Commodities 68*. New York and Geneva: United Nations. https://perma.cc/PMX5-QAML

OECD (n.d.) 'Agricultural Support'. https://doi.org/10.1787/6ea85c58-en

Office of the United States Trade Representative (n.d.) 'Morocco Free Trade Agreement', Executive Office of the President. https://perma.cc/S8KG-K8XV

Partnership on Food and Nutrition Security and Sustainable Agriculture (FNSSA) (n.d.) European Commission, Research and innovation. https://perma.cc/X4AP-UMV4

Pasco, Richard (2010) 'AGOA Countries: Challenges and Considerations in Exporting Horticultural Products to the United States', International Food & Agricultural Trade Policy Council. https://perma.cc/X8G3-KS2M

Peschard, Karine; Golay, Christophe; and Araya, Lulbahri (2023) 'The Right to Seeds in Africa', The Geneva Academy of International Humanitarian Law and Human Rights. https://perma.cc/4ZNK-Q3Q6

Regulation on Deforestation-Free Products (n.d.) European Commission, Energy, Climate change, Environment. https://perma.cc/XKA7-TVBS

Sambidge, Andy (2024) *UAE in talks to buy more African land to aid food security*, AGBI Arabian Gulf Business Insight. https://perma.cc/6PLM-EUP3

Santos, Edna (2016) *Brazil trains African countries on no-till cotton farming*, Embrapa. https://perma.cc/H6FU-4RKE

Schneidman, Witney; McNulty, Kate; and Dicharry, Natalie (2021) 'How the Biden Administration Can Make AGOA More Effective', Brookings. https://perma.cc/W86R-Q7QU

Shi, Ting-Shuai; Collins, Scott L.; Yu, Kailiang; Peñuelas, Josep;
    Sardans, Jordi; Li, Hailing and Ye, Jian-Sheng (2024) 'A global
    meta-analysis on the effects of organic and inorganic fertilization on
    grasslands and croplands', *Nature Communications*, 15(1), p.3411.
    https://doi.org/10.1038/s41467-024-47829-w.

UNCTAD (n.d.) 'UNCTAD|STAT'. UNCTAD. https://perma.cc/A7QD-9ZHM

UPOV International Union for the Protection of New Varieties of Plants
    (2009) 'Explanatory Notes on Exceptions to the Breeder's Right under the
    1991 Act of the UPOV Convention', UPOV. https://perma.cc/63UK-LK6C

USAID (n.d.) 'Feed the Future'. https://perma.cc/E4YN-UMZV

Van Berkum, Siemen (2013) 'Trade Effects of the EU-Morocco
    Association Agreement: Impacts on Horticultural Markets of the
    2012 Amendments', Wageningen, LEI Wageningen UR (University &
    Research Centre), LEI Report 2013-070.
    https://www.researchgate.net/publication/283417560_Trade_effects_of
    _the_EU-Morocco_Association_Agreement

Van der Ven, Colette; and Luke, David (2023) 'Africa in the World Trade
    Organization', in Luke, David (ed.) *How Africa Trades*, London: LSE
    Press, pp.117–40. https://doi.org/10.31389/lsepress.hat.e

WFP (World Food Programme) (2023) 'Impacts of the Suspension of
    the Black Sea Grain Initiative in Eastern Africa', World Food
    Programme, Regional Bureau for Eastern Africa. August 2023.
    https://fscluster.org/sites/default/files/documents/wfp_impacts_of_the
    _suspension_of_the_black_sea_grain_initiative_in_eastern_africa
    _august_2023.pdf

Wintour, Patrick (2023) 'What Was the Black Sea Grain Deal and Why Did
    It Collapse?', *The Guardian*, 20 July. https://perma.cc/29EZ-SQTS

World Bank (2024) 'GDP Deflator (Base Year Varies by Country) | Data',
    World Bank. https://data.worldbank.org/indicator/NY.GDP.DEFL.ZS

World Bank (n.d.) 'World Bank Country and Lending Groups', The World
    Bank. https://perma.cc/V7Z8-D64X

World Meteorological Organization (2023) 'Africa Suffers Disproportion-
    ately from Climate Change', *Reliefweb*. https://perma.cc/FM9P-VMBD

World Trade Organization (2004) 'Agriculture Negotiations Backgrounder:
    Market Access: Special Agricultural Safeguards (SSGs)', WTO.
    https://perma.cc/4CNB-ELS8

WTO (2019) 'Trade Policy Review European Union'.
    https://docs.wto.org/dol2fe/Pages/FE_Search/FE_S_S009-DP.aspx
    ?language=E&CatalogueIdList=297988,293510,264817,259629,239412

,236346,135320,132129,121080,120047&CurrentCatalogueIdIndex=3
&FullTextHash=&HasEnglishRecord=True&HasFrenchRecord
=True&HasSpanishRecord=True

WTO (2023) 'Trade Policy Review: European Union. Report by the
Secretariat', WTO. https://perma.cc/NCY3-BYKS

Yakovenko, M. (2019) 'The Russia–Africa Agriculture Forum: a New
Venue for Cooperation', Russian International Affairs Council (RIAC).
https://perma.cc/24PK-NU2W

9. The World Trade Organization's legal framework and Africa's food security

Colette Van der Ven and David Luke

This chapter undertakes an assessment of the World Trade Organization (WTO) legal framework in relation to food security in Africa. It begins by positing an often-overlooked paradox: the contradictory role that food security plays in international trade. Countries with the means to subsidise production provide food not only for domestic consumption but also for trading in open markets or for giving away as food aid. This enhances global food availability but disincentivises production in poorer and net food-importing countries. Much of the work on food and agriculture in the multilateral trading system is aimed at resolving this conundrum.

After framing the paradox, the chapter reviews the main WTO agreements that impact food security, beginning with the Agreement on Agriculture (AoA). In assessing the AoA, its provisions relating to domestic support, the public stockholding of food supplies, the special safeguard mechanism, and export restrictions are of particular focus. Concerning the last of these, the chapter unpacks the measures taken by WTO member states against the background of the food crisis that followed the Covid-19 pandemic and Russia's invasion of Ukraine. Subsequently, the Sanitary and Phytosanitary (SPS) Agreement, the Fisheries Subsidies Agreement (FSA) and provisions relating to technology transfer from a food security viewpoint are discussed. In each case, we highlight the relevant WTO rules and ongoing initiatives to illustrate the implications for African countries' food security objectives. In particular, the chapter suggests reform initiatives that could be taken by the WTO African Group, the body that coordinates activities among African member states. Finally, in keeping with the aim of this book to put the spotlight on the interrelationship between food trade, food security and climate, the chapter reviews current (at the time of writing) environmental initiatives at the WTO such as the Trade and Environmental Sustainability Structured Discussions (TESSD), which includes environmentally harmful subsidies, with agricultural subsidies among them.

How to cite this book chapter:

Van der Ven, Colette and Luke, David (2025) 'The World Trade Organization's legal framework and Africa's food security', in: Luke, David (ed) *How Africa Eats: Trade, Food Security and Climate Risks*, London: LSE Press, pp. 213–242. https://doi.org/10.31389/lsepress.hae.i License: CC-BY-NC 4.0

9.1 Food security and WTO rules: an unresolved conundrum

Food is unlike any other commodity that is traded. Demand for staples is inelastic. Food price movements are politically sensitive. Food is both traded in international markets and highly protected in domestic markets. During the Uruguay Round (1986–1994) that led to the establishment of the WTO, the negotiators readily accepted that international trade could have both positive and negative implications for food security and crafted a special set of rules set out in the AoA to reflect this reality. On the positive side, the well-known role that trade could play in generating a supply response where food is needed and in emergencies was acknowledged by Uruguay Round negotiators. So also was the role of trade in facilitating access to inputs and modern agricultural technology and infrastructure needed to support agricultural productivity. Today accessing technology through trade is seen as vital to render agricultural production less vulnerable to climate shocks and to drive mitigation and adaptation initiatives.

However, the Uruguay Round negotiators further recognised that, while subsidies and some degree of protection could help to boost domestic food production and safeguard food stability, they also generate trade distortions through overproduction, dumping on world markets, price depression and the destabilisation of local production. To curtail the risks, negotiators agreed to allow subsidies and protection within disciplines that were laid out in the AoA. Net food-importing countries were acknowledged to be especially vulnerable to disruptions in global food supply chains. This category of countries was singled out for special consideration in further deliberations on food and agriculture. To make the point, a Decision on Measures Concerning the Possible Negative Effects of the Reform Programme on Least-Developed and Net Food-Importing Developing Countries (NFIDCs) was adopted (Decision on Measures Concerning the Possible Negative Effects of the Reform Programme on Least-Developed and Net Food-Importing Developing Countries, 1994). The decision set out pathways for mitigating negative outcomes for least-developed countries (LDCs) and NFIDCs – which include 42 African countries – such as food aid, technical and financial assistance and special conditions for agricultural export credit disciplines. In addition, food security as an objective is explicitly mentioned in specific WTO agreements such as the General Agreement on Tariffs and Trade 1994 and the AoA, which were part of the Uruguay Round trade deal.

Following the formation of the WTO in 1995, food security has remained central to negotiations on agriculture. In the decades since, especially over the last decade and most recently in 2022 at the Twelfth Ministerial Conference, the 164 Members (as WTO member states are known) delivered a series of outcomes that are in line with the obligations of the NFIDC Decision (World Trade Organization 2024a). But how to discipline agricultural subsidies, afford some level of protection for NFIDCs and LDCs, and keep world food markets open is a conundrum that remains unresolved at the WTO, as will be seen in this chapter.

9.2 The Agreement on Agriculture

The AoA comprises three main pillars: market access, domestic support and export competition. Under the market access pillar, Members were required to replace agriculture-specific non-tariff measures with a tariff that afforded equivalent levels of protection. This is also known as tariffication. Quantifying the amount of protection Members were providing was the first step towards implementing reduction targets. Mandatory minimum and average tariff reduction requirements were established for developed and developing countries. LDCs were required to bind agricultural tariffs but were exempt from undertaking tariff reductions (World Trade Organization 2024b).

The pillar on domestic support focuses on the use of subsidies and other support programmes that directly stimulate agriculture production. These provisions seek to discipline the use of domestic support, while at the same time leave room for governments to design agricultural policies. The rules reflect a conceptual distinction between two types of domestic support: subsidies that provide minimal or no trade-distortive effect, and subsidies that are trade-distortive. Members were required to make annual reductions to the latter category of subsidies: by 20 per cent over a six-year implementation period for developed countries, and 13.2 per cent over a 10-year period for developing countries. Annex 2 and Annex 3 of the AoA set out, respectively, domestic support measures that are exempt from reduction requirements and the rules to calculate domestic support (World Trade Organization 2024c).

The export competition pillar covers the use of export subsidies and other government support programmes that subsidise exports. Export subsidies on agricultural products are permitted for those WTO Members that reserved this right in their schedule of concessions but subject to reduction commitments (World Trade Organization 2024d). However, the trade-distorting effects of export subsidies became a prime target for criticism by civil society and other stakeholders. In 2015, at the WTO Ministerial Conference in Nairobi, Kenya – the first time this had been held in an African country – Members agreed that developed and developing country Members must eliminate the remaining scheduled export subsidy entitlements within specified timeframes (Tenth WTO Ministerial Conference, Nairobi, 2015 2024).

For the purposes of our discussion in this chapter, we will mainly focus on the domestic support pillar.

9.3 Agricultural subsidies

An overview of subsidy disciplines

Agriculture is widely subsidised because of its food security implications. These subsidies amount to hundreds of billions of dollars each year. The United States, China, the European Union (EU) and India top the list of subsidisers. In 2019–2020, the United States provided €190.6 billion in domestic support,

China provided €173.1 billion, the EU €81 billion and India €67.7 billion (Directorate-General for Agriculture and Rural Development n.d.). These subsidies potentially create market distortions that negatively impact agricultural producers that do not benefit from such generous subsidies.

The AoA disciplines only agricultural subsidies that are trade-distortive. To differentiate between different types of agricultural subsidies, the AoA categorises agricultural subsidies into four boxes: the Amber Box, the Blue Box, the Green Box and the Development Box. Only subsidies that fall into the Amber Box are subject to reduction requirements, set out in WTO Members' schedules. The differences between the boxes are explained below.

The Amber Box covers the most trade-distorting subsidies, which are subject to limitations based on the country's Final Bound Total Aggregate Measurement of Support (FBTAMS) entitlements.[1] Examples of these subsides include price support regimes that regulate prices and production amounts; systems or targets for minimum prices for agricultural commodities; and highly subsidised insurance schemes and other forms of protection for farmers against low yields (Lau and van der Ven 2017).

For trade-distortive subsidies that fall into the Amber Box, WTO Members were allocated different levels of aggregate measure of support (AMS) entitlement. Using 1986–1988 as the base period, developed countries that were subsidising agriculture during this period had to reduce the level of support by 20 per cent over six years and developing countries by 13 per cent over 10 years – expressed in terms of total AMS (World Trade Organization 2024c). Only 33 WTO Members enjoy FBTAMS entitlements (see Table 9.1). The WTO Members that are not included on this list did not subsidise their agricultural sector during the base period of 1986–1988. Accordingly, these countries were not allocated an FBTAMS entitlement.

The amount of trade-distorting domestic support any WTO Member can provide, irrespective of their FBTAMS entitlements, is also determined by *de minimis* thresholds, that is, a percentage of the value of production that does not need to be counted towards a WTO Member's FBTAMS entitlements. These percentages differ based on a country's development status: for developed countries it is 5 per cent, and for developing countries it is 10 per cent.[2] Importantly, WTO Members are permitted to provide product-specific support under the *de minimis* provisions. For developing countries and LDCs that have no FBTAMS entitlements, the *de minimis* allowance is critical.

The Blue Box covers subsidies that may have some trade-distortive effects by limiting production or establishing production quotas, or payments to farmers for repurposing farmland. Blue Box subsidies are not counted towards a Member's AMS entitlements. An example of a Blue Box subsidy is US payments to farmers who participate in its Acreage Reduction Programme, which requires idling of farmland. Blue Box subsidies are hardly used by developing countries as they involve direct payments, which implies significant budgetary outlays. To date, no African country has made use of this type of subsidy.

Table 9.1: WTO Members with FBTAMS reduction commitments under the Amber Box

Argentina	Jordan	**South Africa**
Australia	Korea	Switzerland
Brazil	Mexico	Liechtenstein
Canada	Moldova	Chinese Tapei
Colombia	Montenegro	Tajikistan
Costa Rica	**Morocco**	Thailand
EU	New Zealand	**Tunisia**
North Macedonia	Norway	Ukraine
Iceland	Papua New Guinea	United States
Israel	Russian Federation	Venezuela
Japan	Saudi Arabia	Viet Nam

Source: World Trade Organization (2024e).
Note: African countries are shown in bold.

The *Green Box* covers subsidies that are deemed to be minimally trade-distorting. These subsides are exempt from reduction commitments. Green Box subsidies are listed in Annex 2 to the AoA and include horizontal activities such as research, training and certain types of direct payments to producers not linked to production.[3] Subsidies to achieve environmental objectives such as land rehabilitation, soil conservation, resource management, drought and flood control fall within the Green Box (WTO 2013). These subsidies are required to be provided under a publicly funded programme and do not involve transfers from consumers. As was noted in Chapter 4, the Malabo Declaration requires that African countries allocate at least 10 per cent of public expenditure to agricultural development. This would be covered mainly as part of the Green Box and will not be counted towards AMS limits.

The *Development Box* provides flexibilities for developing countries, by exempting certain types of subsidies they provide from being counted towards a WTO Member's FBTAMS. These subsidies include inputs such as irrigation systems and fertilisers for low-income producers and outlays for the acquisition of machines and provided they are used to promote agricultural and rural development and form an integral part of development programmes.

Agricultural subsidies and implications for food security in Africa

The domestic support disciplines in the AoA have been criticised because of their role in exacerbating structural asymmetries in agricultural subsidies between developing and developed countries. Differences in FBTAMS entitlements, which were calculated based on the domestic support Members

Table 9.2: Distribution of FBAMs entitlements (2018)

WTO Member	FBTAMS entitlement (US$ billion)	Cumulative share of FBTAMS entitlement (%)
EU	81.03	48
Japan	36.45	21.6
US	19.1	11.3
Mexico	12.82	7.6
Others	19.52	11.6
Total	168.92	100

Source: World Trade Organization (2023a).

provided between 1986 and 1988, have led to a situation where the distribution of FBTAMS entitlement is highly skewed. Four WTO Members (the EU, the US, Japan and Mexico) account for 88.4 per cent of FBTAMS entitlements, with the EU alone accounting for 48 per cent (World Trade Organization 2023a) (see Table 9.2).

Only three African countries – South Africa, Morocco and Tunisia – have FBTAMS entitlements.[4] This means that only these countries are permitted to provide Amber Box subsidies up to the limit specified in their schedules. All other African WTO Members can subsidise only within *de minimis* levels (up to 10 per cent of the value of agricultural production). Both Morocco and Tunisia have notified Amber Box use, but not beyond their *de minimis* levels, whereas South Africa has notified zero Amber Box use (World Trade Organization 2023a).

The African Ministers of Trade Declaration on WTO issues, submitted in June 2022, noted that 'long term resilience to future food crises and sustainable food security lies in unlocking the agricultural productive capacity of African economies through addressing longstanding asymmetries and imbalances in the Agreement on Agriculture' (World Trade Organization 2022a). Indeed, it is difficult, especially for small-scale agricultural farmers in Africa, to compete with heavily subsidised agricultural imports. This is especially the case given that most of the support provided goes to five commodities: rice (US$26.5 billion), wheat (US$13.3 billion), dairy (US$10.3 billion), bovine meat (US$8.5 billion) and corn/maize (US$8.3 billion) (World Trade Organization 2023a). Moreover, in spite of possible beneficial effects on food availability, recent studies have found that food subsidies support neither sustainability nor human health, and generate almost US$12 trillion in hidden costs (FAO, UNDP and UNEP 2021; Food and Land Use Coalition 2019).

African WTO Members have historically opposed the use of Amber Box subsidies. Specifically, the African Group has proposed that WTO Members with scheduled FBTAMS entitlements that exceed the *de minimis* levels must apply a cap on their non-product-specific FBTAMS at their *de minimis* level

(African Group 2023). For product-specific support, they have proposed a cap on FBTAMS entitlements at *de minimis* levels for WTO Members that account for a share of 10 per cent or above of all WTO Members' FBTAMS, or account for 8 per cent of global exports of WTO Members within a period of two years and five years for all other WTO Members (African Group 2023).

Reducing or eliminating FBTAMS can generate efficiency gains with positive impacts for African countries. This is because domestic support – which is being provided most generously outside the African continent – can lead to price suppression, thereby disincentivising domestic production. For example, a study found that US subsidies alone depressed global maize prices by about 9 to 10 per cent (Ambaw et al. 2021). Another study found that removing domestic support for cotton globally could result in an increase of the value of net cotton exports by African cotton producers by US$622 million per year (Anderson et al. 2021). It is clear that it would be difficult for African farmers to sell products at the depressed world price without receiving similarly generous domestic support. Removing these trade-distortive, price suppression subsidies by reducing or eliminating FBTAMS allowances could lead to increased food production, including in NFIDCs, and enable African countries to achieve self-sufficiency.

However, removing or reducing FBTAMS can also negatively impact food affordability, globally and in Africa. A reduction of FBTAMS subsidies will likely result in a decline in farm output and subsequent increases in food prices. This could make it more costly for African countries, many of which are NFIDCs, to import the staple commodities crucial to achieve food security. Moreover, a shift towards higher levels of agricultural production in Africa enabled by the reduction or removal of AMS would not occur overnight. From this point of view, advocating for reducing or removing FBTAMS is a conundrum that cannot be easily resolved.

Trade-offs must also be considered with regard to *de minimis* entitlements. Except for the three African countries with FBTAMS allowances, African WTO Members are only entitled to provide trade-distortive support as part of their *de minimis* allowance. The African Group's position is to keep the *de minimis* allowances to ensure policy space for Africa's agricultural development. To the extent that African policymakers consider this desirable, domestic support must be linked to productivity targets to avoid waste and inefficiencies.

Given that *de minimis* is calculated as a percentage of a country's total value of production, *de minimis* allowances have become increasingly large as global agricultural production has increased. Whereas global *de minimis* entitlements were around US$182.4 billion in 2001, it more than tripled to US$631.8 billion in 2019, with China and India in the lead (World Trade Organization 2023a). In fact, China has the most Amber Box entitlements even if it does not have FBTAMS entitlements, owing to its *de minimis* share.[5] Maintaining *de minimis* allowances across the board could have negative implications for African food security with a similar global price depression effect as FBTAMS allowances. At the same time, reducing *de minimis* allowances could lead to a reduction in food being produced globally, with potentially negative implications for the African continent. (It is surprising that there is little available research on the possible global effects of new disciplines on agriculture at the WTO. More research is needed, which will also provide fresh evidence for WTO negotiations.)

The African Group has also proposed that any developing country Member experiencing a severe food crisis should have recourse to product-specific *de minimis* exceeding the 10 per cent of the value of production threshold, provided that the country exports less than 1 per cent of that product globally (African Group 2023). While this would provide African WTO Members with additional policy space to address a severe food crisis, it assumes that restricted policy space is what stands in the way of additional domestic support. However, this is not necessarily the case, given that not all African countries have used their *de minimis* allowance (World Trade Organization 2023a).

Trade-offs may also be required as regards the Development Box and its implications for African food security. Currently, the African Group's position is to keep the Development Box as set out in Article 6.2 of the AoA as is – notwithstanding calls from some WTO Members to impose a cap. Under the Development Box, developing country WTO Members can provide a variety of subsidies to develop agricultural production. However, the notifications of the Development Box suggest that African WTO Members are only marginal users of the Development Box.[6] Asia is responsible for at least 85 per cent of usage in most years between 2001 and 2019, with India topping the list at the forefront of Article 6.2 expenditures (Committee on Agriculture 2021a). When advocating to maintain Development Box privileges, African countries should consider whether they have the financial means to effectively use the policy space it provides and be mindful of how the use of these flexibilities by other developing countries could negatively impact African agriculture production (Ambaw et al. 2021). Perhaps the African Group could explore limiting Development Box benefits only to NFIDCs and LDCs.

Concerning support provided under the Green Box, an African Group communication noted an emerging trend whereby measures that are notified as Green Box support by developed countries under the AoA paragraphs 5–15 of Annex 2 are not decoupled from production (African Group 2023). This is known as box shifting, using the Green Box for Amber Box measures. Since support provided under the Green Box does not need to be counted towards a Member's FBTAMS, some WTO Members increase their subsidy allowances through box shifting. In 2020, 28 WTO Members provided Annex 2 support above US$100 million, with 12 of these WTO Members providing Annex 2 support that exceeded 5 per cent of the annual value of production, and nine exceeding 10 per cent of the value of production (African Group 2023). Similar to the FBTAMS, this can have negative implications for African producers. To avoid the trade-distorting effect of Green Box support, the African Group has proposed to introduce a cap – at 5 per cent of the value of production – with exemptions for farmers with low-income levels in developing countries and LDCs (African Group 2023).

9.4 Public stockholding of food supplies

Public stockholding programmes are policy tools used by governments to purchase, stockpile and distribute food when needed (World Trade Organization 2024f). Specifically, public stockholding programmes provide (1) emergency stocks to reduce the vulnerability of consumers to supply disruptions or food price shocks in emergencies; (2) buffer stocks to stabilise prices within the domestic market to avoid excessive volatility; and (3) stocks for domestic food distribution or for external food aid (Avesani 2023). Most African WTO Members, and some RECs like ECOWAS, have public stockholding programmes in place. However, despite the prevalence of public stockholding programmes, they might not be sufficient to address emergency situations given low stock-to-use ratios in many African countries (Gro Intelligence n.d.).

While public stockholding programmes are essential for food security, they are disciplined by the AoA for their potential to distort market prices and trade. The AoA allows governments to procure stocks at current market prices. However, if stocks are procured at pre-announced administered prices, outlays are counted towards a country's AMS, owing to its potential market-distortive effects (Sinha and Glauber 2021). Developing countries, including African WTO Members, have raised concerns that procurement at administered prices could push them towards exceeding allowable limits, thus limiting their ability to pursue public stockholding programmes to meet their food security needs.

Following the launch of the Doha Round negotiations, the African Group advocated for the removal of references to AMS with respect to public stockholding programmes, effectively seeking to put these programmes in the Green Box category (World Trade Organization 2014). In 2013, at the Ninth Ministerial Conference in Bali, WTO Members reached an interim solution for stockholding known as the 'Peace Clause'. Under the 'Peace Clause' WTO Members agreed to refrain from challenging food security programmes of developing countries that exceeded *de minimis* or bound limits provided certain transparency conditions were met (World Trade Organization 2013). Following the Nairobi Ministerial Conference two years later, the WTO General Council adopted a decision to extend the Peace Clause indefinitely while continuing to work towards a permanent solution (WTO General Council 2014).

In March 2020, India became the first WTO Member to invoke the Bali Decision on Public Stockholding when it notified the organisation that it had exceeded its *de minimis* support level for rice as a result of its minimum support price programme and other welfare schemes (World Trade Organization 2020). India submitted similar notifications on breaching its permitted support levels for rice in 2021, 2022 and 2023 (Committee on Agriculture 2021b;

Committee on Agriculture 2022a; Committee on Agriculture 2023a). Yet, following these notifications, some WTO Members raised questions about a surge in India's rice exports.[7] India's experience highlights the limitations and stringent requirements that WTO Members are subjected to when invoking the Peace Clause.

The African Group, the African, Caribbean and Pacific (ACP) Group and the G33 (a group of developing countries that includes India and China) sought to address shortcomings perceived in the Bali Decision. First, the Bali Decision applied only to programmes that existed at the time the decision was taken but not to new or future programmes. This limited the scope of the Peace Clause (Matthews 2014).[8] Second, the Bali Decision can be utilised only for 'traditional staple food crops', i.e. primary agricultural products that are predominant staples in the traditional diet of a developing Member (World Trade Organization 2014, footnote 2 to para. 2). In other words, the Bali Decision does not apply to agricultural commodities that are not part of a developing Member's traditional diet. This creates a narrow group of products that are eligible for the Peace Clause than that which may be eligible for public stockholding. Third, the Bali Decision sets out onerous notification requirements that must be met to benefit from the Peace Clause.

The proposal submitted by African Group, ACP and G33 sought to address these limitations and included a permanent solution for public stockholding. The proposal aimed to recalculate trade-distorting support when stocks are procured at administered prices, adjusting for excessive inflation. For situations in which a developing country WTO Member exceeds its allowable support because of public stockholding programmes for food security purposes, the proposal sets out anti-circumvention measures that aim to buffer potential market-distortive effects. These include ensuring that stocks acquired under the public stockholding programme for food security purposes do not adversely affect the food security of other Members and a best endeavour provision to refrain from exporting stocks acquired through public stockholding programmes, except in situations of international food aid or when requested by net food-importing developing countries or similar situations of food scarcity (African Group, ACP and G33 2022). Finally, the proposal provides recommendations for less onerous transparency and notification requirements than under the Bali Decision.

Carving out additional policy space for public stockholding programmes through a permanent solution, as proposed by the African Group, ACP and G33, would enable African WTO Members to ensure food availability for critical crops during times of food scarcity. At the same time, the proposal for a Permanent Decision on Public Stockholding also opens the door to distorting global agricultural markets even further, as it would allow developing countries with large agricultural markets, like China and India, to provide unlimited support. While this could negatively impact food production in Africa due to price suppression, it could at the same help to ensure that, globally, sufficient food is being produced.

To better balance the dual objectives of providing policy flexibility to advance food security in developing countries while preventing trade distortion, one solution could be to limit outcomes on a permanent public stockholding programme to LDCs and NFIDCs, or to WTO Members whose procured stocks do not exceed a certain percentage of the average value of production (Ambaw et al. 2021), or whose share in world trade in agricultural products amounts to no more than a set percentage (Avesani 2023). For these countries, the risk that public stockholding programmes would result in global market distortions is generally low, given that many LDCs do not have the financial capacity to procure food at administered prices (WTO General Council 2014). Politically, however, doing this will be very challenging, given that the G33 is one of the staunchest proponents of a permanent solution on public stockholding. Complicating this further is the fact that a permanent solution has been elusive, and WTO Members have not made any progress since the Bali Ministerial Conference a decade ago.

9.5 Special safeguard mechanism

The AoA includes a special agricultural safeguard provision (SSG), but its applicability is limited to the 39 countries that undertook tariffication[9] of agricultural products during the Uruguay Round. This included the following African countries: Botswana, Eswatini, Morocco, Namibia, South Africa and Tunisia.[10] However, WTO Members that did not engage in tariffication during the Uruguay Round are not eligible to use the SSG. Establishing a special safeguard mechanism (SSM) with broader eligibility would address this gap. At the 2015 Nairobi Ministerial Conference, WTO Members adopted a decision to negotiate an SSM for developing countries to enable them to temporarily increase tariffs on agriculture products in cases of import surges or price declines (World Trade Organization 2015).

It is vital for Africa to protect its resource-poor and small-scale farmers from excessive price volatilities in agriculture commodities. While the dynamics on price volatility for rural African households are complex (G33 2017), price falls coupled with import surges are especially problematic as farmers risk losing expected returns, which could take them further into poverty. In a 2019 proposal to the WTO, the African Group noted that African countries have been 'subject to massive and repetitive import surges, resulting over the years and in the absence of any means to safeguard the market from substantial reduction in production amounting in some cases to more than 50 per cent decrease, and the loss of numerous jobs' (Benin on behalf of the African Group 2019). A 2020 study by Das of eight developing countries,[11] Ghana, India, Indonesia, Namibia, Philippines, Senegal, Sri Lanka and Turkey, showed that these countries experienced import surges covering between 191 and 348 tariff lines (Das et al. 2020). Given Africa's prevalence of smallholder farmers, minimising the impact of a commodity international price collapse on domestic prices is critical.

However, the Twelfth Ministerial Conference, in 2022, made no progress on the SSM, reflecting the sensitivity of this issue and disagreement among the Members. While developing countries are pressing for the SSM, other Members have sought to ensure that the SSM is limited by discipline in order not to compromise market access reform efforts in the negotiations of existing tariff bindings. Concerns were also raised on the potential negative implications of the SSM on Members' exports, and on trade between WTO Members more broadly.

Nonetheless, the LDC Group, the African Group and the G33 have called for the adoption of a simple and accessible SSM to be used as a trade remedy tool to balance distortions in agricultural markets. The African Group called upon Members to intensify discussions on SSM to reach an outcome at the Thirteenth Ministerial Conference, held in February 2023. Its proposal advocated for an SSM that would cover both price and volume-based triggers with no *a priori* product limitations as to its availability, and one that would be easy to apply by developing countries (Committee on Agriculture 2023b). The African Group further proposed that any transparency requirements should not be excessively onerous for developing countries (Committee on Agriculture 2023b). The African Group requested a moratorium on Members from challenging the compliance of a developing country Member with its SSM obligations through WTO dispute settlement mechanism pending the entry into force of a potential SSM-related amendment or protocol to the AoA (Committee on Agriculture 2023b). However, no decision on an SSM was taken at the Thirteenth Ministerial Conference, which was deadlocked on most issues on its agenda.

As African Members advance a food security agenda at the WTO, it will be important to continue to press for an SSM. Meanwhile, in situations marked by price volatility, African Members should also consider the extent to which existing tariffs could be applied to protect vulnerable smallholder farmers from import surges. The 2020 study by Das et al. referenced earlier found that the countries studied had differences between the applied and bound levels in their tariff schedules of over 20 per cent, suggesting that simply raising the tariff up to the bound level could be another method to protect against import surges (Das et al. 2020). Until a SSM has been negotiated, African WTO Members, which tend to have high bound tariffs for agricultural products, should consider to extent to which existing tariff schedules could provide a temporary buffer.

9.6 Export restrictions on agricultural products

As most African countries are net food importers, disruptions in food supply chains can be catastrophic. As discussed in earlier chapters, export restrictions contributed to increased price volatility and higher price levels during the Covid-19 pandemic and following Russia's invasion of Ukraine (Food and

Agriculture Organization of the United Nations, World Trade Organization and World Bank 2023). This followed similar episodes in 2008–2010 and the 1970s (Giordani, Rocha and Ruta 2012; Trade and Markets Division 2009, p.11).

Export restrictions can temper domestic price increases, or ensure sufficient domestic supply is available in case of scarcity. This is especially critical in situations of food shortage. However, export restrictions also accelerate price spikes in international markets, and can have a broader destabilising effect on global markets as trade is interrupted abruptly. This has direct implications for the availability and affordability of food in domestic markets (Committee on Agriculture 2023c). In the first six months of 2022, countries adopted 75 export restrictions affecting trade in food and fertiliser (Espitia, Rocha and Ruta 2022). Export bans on rice, wheat and citrus fruits, including by major exporters such as India, Russia and Turkey, led to price increases estimated at 12.3 per cent, 9 per cent and 8.9 per cent, respectively. During the same period, export prices for soya bean oil and maize increased by 14 and 6.1 per cent, respectively (Espitia, Rocha and Ruta 2022).

These price increases were challenging for NFIDCs and LDCs. In June 2022, 26.3 per cent of LDC agricultural imports (measured in calories) were impacted by export restrictions, compared to 13.2 per cent for developed countries (Glauber et al. 2022). Moreover, many LDCs do not have the financial resources to compete for access to alternative markets at higher prices (Committee on Agriculture 2023c), and so experience higher levels of food inflation as a result of the supply shortages (Committee on Agriculture 2023c).

Article XI of the General Agreement on Tariffs and Trade (GATT) 1994 disciplines the adoption of quantitative restrictions. Generally, it prohibits quantitative restrictions, 'whether made effective through quotas, import or export licenses or other measures', but permits the use of export restrictions to relieve critical food shortages. Thus, export restrictions that were adopted in response to the Covid-19 pandemic and the Russian aggression on Ukraine were, to the extent they were necessary to relieve critical food shortages, not considered WTO-inconsistent.

Article 12 of the AoA added a transparency requirement in making it mandatory for WTO Members that are net food exporters and which adopt a food export prohibition or restriction to (1) give due consideration to the effects of such prohibition or restriction on importing Members' food security; and (2) give notice in writing to the Committee on Agriculture. These transparency requirements do not, however, apply to inputs such as fertilisers (Calvo 2023).

At the Twelfth Ministerial Conference, in 2022, which was held during a period of exceptional turbulence in world food markets, WTO Members sought to further discipline export restrictions by adopting two ministerial declarations: the Ministerial Declaration on the Emergency Response to Food Insecurity (WTO Food Security Declaration) and the Ministerial Decision on World Food Programme Food Purchases Exemption from Export Prohibitions or Restrictions (Ministerial Decision on WFP Exemptions).

The WTO Food Security Declaration includes a provision to 'ensure that any emergency measures introduced to address food security concerns shall minimize trade distortions as far as possible; be temporary, targeted, and transparent; and be notified and implemented in accordance with the WTO rules' (Ministerial Declaration on the Emergency Response to Food Insecurity (WT/MIN (22)/28), 2022, art 5). A related provision in the WTO Food Security Declaration is summarised in Box 9.1 below. While the declaration should be applauded for seeking to minimise trade disruptions on food products caused by export bans, it is not likely to lead to significant changes in WTO Members' behaviour vis-à-vis export restrictions on food, given that it does not establish any new binding rules against export restrictions on food (Calvo 2022a).

The Ministerial Decision on WFP Exemptions provides that Members shall exempt foodstuffs purchased for non-commercial humanitarian purposes by the World Food Programme (WFP) from export prohibitions or restrictions to ensure the steady supply of its humanitarian aid (World Trade Organization 2022b). Given that the WFP is a humanitarian organisation that delivers food assistance in emergencies, including in many African countries, this ministerial decision could help to ensure the WFP's access to available food supplies.[12] However, the decision also underlined that its provisions 'shall

Box 9.1: A dedicated WTO work programme on food security

Another aspect of the 2022 WTO Decision on Food Security of interest to Africa was the establishment of a dedicated work programme to consider the needs of LDCs and NFIDCs in increasing their resilience, bolstering domestic production, and enhancing their domestic food security. In line with this mandate, a work programme under the Committee on Agriculture was established by the Members in November 2022 with four thematic areas: (1) access to international food markets; (2) financing of food imports; (3) agricultural production and resilience of least-developed and net food-importing developing countries; and (4) horizontal issues.

Technology transfer and knowledge cooperation on climate resilient agriculture development and coordinated rapid response in case of food security crises are some of the areas identified for further discussion. As a first step, the work programme issued a questionnaire to identify the utilisation of WTO flexibilities by least-developed and net food-importing developing countries.

Source: Committee on Agriculture(2022b). Also see Committee on Agriculture (2022c).

not be construed to prevent the adoption by any Member of measures to ensure its domestic food security in accordance with the relevant provisions of the WTO agreements' (Reuters 2022). In effect the decision sought to balance the WFP exemption and a WTO Member's ability to adopt measures to ensure its own food security. While this may appear contradictory, implementation of the decision in good faith by WTO Members might help to tackle the food crisis during the early years of the 2020s by ensuring that critical relief reaches the most vulnerable.

In the absence of clearer and binding disciplines for food export restrictions and prohibitions, the 2022 measures signal a desire among WTO members to cooperate to ensure that vulnerabilities are not left unaddressed. This is also an opportunity for WTO Members to deliver on additional outcomes for clarifying export restrictions and disciplines on prohibitions in relation to both Article XI of GATT 1994, and Article 12 of the AoA. Moreover, notifications of export restrictions are still lacking. There is scope for the African Group to call for more transparency on export restrictions notifications. The African Group could also seek exemptions from export restrictions or prohibitions for food destined to LDCs and NFIDCs in periods of acute food instability (Committee on Agriculture 2023d).

9.7 The Sanitary and Phytosanitary Agreement

Food security cannot be achieved without access to safe food and inputs like seeds. Food standards and trade mutually contribute towards delivering safe, nutritious and sufficient food for the world's population. On the other hand, foodborne diseases contribute to the incidence of malnutrition and erode food security. The 2015 WHO Estimates of the Global Burden of Foodborne Diseases report estimated that in Africa food safety hazards were responsible for approximately 137,000 annual deaths and about 91 million cases of acute foodborne illnesses, the highest estimates worldwide (World Health Organization 2015). The economic burden as a result of productivity loss associated with foodborne diseases in low- and middle-income countries was estimated at US$95.2 billion per year in 2019 (Jaffee et al. 2019).

Food safety falls under the ambit of the WTO Sanitary and Phytosanitary Agreement, which allows WTO Members to set their own standards on food safety and plant and animal health but puts a premium on measures that are based on mutually agreed international standards.

At the 2022 WTO Ministerial Conference, Members agreed on an SPS Declaration to enhance the implementation of the Sanitary and Phytosanitary Agreement and manage issues related to international trade in food, animal and plants (World Trade Organization 2022d). The declaration specifically identifies 'climate change and increasing environmental challenges and associated stresses on food production' as a challenge and the 'growing importance of sustainable agricultural practices and production systems, including their contribution to addressing climate change and biodiversity conservation' as one of the opportunities for addressing emerging challenges. It provides for the establishment of a work programme to explore how the Sanitary and Phytosanitary Agreement can contribute to global food security and sustainable food systems, enhance safe international trade in food through adaptation of measures to regional conditions, and address the needs of developing and least-developed Members in the elaboration and application of SPS measures, among others. The African Group was supportive of this ministerial declaration during the negotiations and can benefit from discussions that emanate from the working group (World Trade Organization 2022a). It would be important for the group to play an active role in the work programme discussions to ensure that it addresses Africa-specific SPS issues.

9.8 The Fisheries Subsidies Agreement

Seven million tons of fish are caught annually in Africa (African Development Bank 2022) and over 12 million people in Africa depend directly or indirectly on the marine fishing industry for their livelihoods (World Trade Organization 2023b). Fish is also critical for Africa's food security as an important protein source for over 400 million Africans, as discussed in Chapter 3. According to forecasts, the continent must produce an additional 1.6–2.6 million tons of fish a year by 2030 to meet consumption needs (Fevrier and Dugal 2017). The African Union considers the fisheries sector to be 'Africa's future', highlighting the sector's role as a 'catalyst for socio-economic transformation' (World Bank and United Nations Department of Economic and Social Affairs 2017). Fishery around African coasts and islands is mainly artisanal and generally carried out through traditional practices that are sustainable. Foreign subsidised commercial fleets dominate both national territorial waters beyond the coasts and the high seas around the continent. Much of this is illegal, unreported and unregulated (IUU) fishing. It generates as much as US$2.3 billion in lost revenue to African countries each year and leaves more than 30 per cent of African fish stocks overfished (World Trade Organization 2023c).

The Fisheries Subsidies Agreement (FSA) was adopted at the 2022 Ministerial Conference to discipline IUU practices. It prohibits IUU as well as subsidies to fishing overfished stock and subsidies to fishing on the unregulated high seas. The FSA includes reporting and notification obligations and provides flexibilities for developing countries and LDCs with regard to some of the obligations.

The challenges caused by subsidised foreign fleets in Africa suggest that these disciplines could be highly beneficial to African food security (World Trade Organization 2023c). Curtailing capacity-enhancing subsidies could reduce overcapacity and the ability of foreign fleets to exploit Africa's fishery resources (African Development Bank 2022). But many African countries do not have the required resources for monitoring fisheries activities through patrolling and inspections at sea. Ahead of the adoption of the FSA, African Ministers of Trade emphasised that 'an outcome on the fisheries subsidies negotiations must not undermine the right of coastal states and fully respect their territorial integrity and sovereignty' (African Ministers of Trade 2022).

African WTO Members also expressed concern that the FSA should not reduce the available policy options to further develop domestic fishery sectors. During the FSA negotiations, the African Group stressed the importance of special and differential treatment (SDT) – reflecting common but differentiated responsibilities and respective capabilities under the Paris Agreement – to ensure food security and protect the livelihoods of coastal communities, as well as *de minimis* threshold to exempt artisanal and small-scale fisheries. While the FSA includes SDT provisions, it does not fully exempt developing countries and LDCs or artisanal and small-scale fisheries from the disciplines it sets out.

However, the FSA addresses the resource and capacity constraint of poor countries. It envisages the creation of a WTO funding mechanism to provide targeted technical and capacity-building assistance to help integrate sustainability elements into fisheries policies and practices, strengthen sustainable fisheries management systems, and comply with notifications and transparency obligations.

There is some evidence that African countries are leading beneficiaries of official development assistance, totalling 48 per cent of all fisheries disbursements. Mozambique, Madagascar, Nigeria, Tanzania and Senegal topped the list of countries receiving funds for sustainable ocean economy initiatives between 2010 and 2020 (World Trade Organization 2022c). It would be critical for African Members to identify the specific types of support they require to advance sustainable fishery management practices and comply with the reporting requirements set out in the FSA. This will include strengthening African governments' sea patrolling capacity, as well as evidence and data collection (Walker, Reva and Willima 2022).

At the time of writing, WTO Members are negotiating outstanding issues such as regulating subsidies that promote overfishing and overcapacity. Adopting additional disciplines on overfishing and overcapacity would be of interest to Africa, not only from a food security perspective but also in view of the lost revenues, estimated to be around US$2.3 billion annually (World Trade Organization 2023c). African Members also remain concerned that FSA disciplines limit their policy options to provide support to small-scale artisanal fishing and seek appropriate exemptions (World Trade Organization 2023d). To this end, in current and future negotiations, African Members could propose exemptions based on a *de minimis* threshold, measured in percentage of a WTO Member's global fish stock. The FSA uses 0.8 per cent annual share of the global volume of marine catch as the threshold for notification obligations. This figure could also be used as a *de minimis* threshold on which agreement might be easier to reach than seeking exemptions for artisanal fisheries, taking also into account the difficulty in reaching a consensus on the definition of artisanal fishing.

9.9 Technology transfer and food security

As discussed in earlier chapters, low agricultural productivity and low yields are ubiquitous problems across Africa. An important means of overcoming this is the use of technology including hybrid seeds, fertilisers, pesticides, mechanical equipment, and veterinary care for livestock and poultry (General Council, Committee on Agriculture and Committee on Trade and Development 2023). Adopting these and other smart agricultural technologies would not only increase agricultural yields but also support African agriculture to adapt to climate change and extreme weather occurrences such as floods and droughts.

The importance of facilitating access to smart agricultural technologies has been recognised at the WTO. Responses to a questionnaire survey of NFIDC and LDC Members that was discussed in the Committee on Agriculture revealed that access to inputs, agricultural equipment, capacities and support for absorbing new agricultural technologies are priorities (Committee on Agriculture 2023d). Other issues that were highlighted were early-warning systems, storage and supply-chain infrastructure to contain food losses, regulatory infrastructure for SPS, high-yielding seeds and livestock breeds, and 'assistance to promote diversification of production and production of nutritious local products entailing financial prudence and sound environmental practices' (Committee on Agriculture 2023d).

In July 2023, the African Group circulated a communication on the role of transfer of technology to build agricultural resilience (General Council, Committee on Agriculture and Committee on Trade and Development 2023). The communication noted that:

> Effective technology transfer also holds the potential to contribute to addressing the risks of concentration of production and supply of agri-food products which evidently renders import-dependent countries vulnerable to global supply chain shocks. It can therefore contribute to building resilience, especially developing countries, including least-developed countries and net food importing countries, address food insecurity, and support initiatives towards more environmentally sustainable farming methods in light of the climate change challenge. (General Council, Committee on Agriculture and Committee on Trade and Development 2023)

For patented technologies, African WTO Members could seek to utilise the provisions on compulsory licensing as set out in Article 31 of the Trade-Related Aspects of Intellectual Property Rights (TRIPS) Agreement. Owing to the built-in limitations of these provisions, African countries might want to seek clarifications on the applicability of compulsory licensing to smart agricultural technologies. However, even if technologies critical for agricultural resilience could be exempted from the limitations on compulsory licensing, having access to patents must be coupled with adequate technological capacity and specific know-how of the production process in order to develop the product.

Another TRIPS provision that will be useful to ensure African LDCs are able to access smart agricultural technologies is Article 66.2, which requires developed countries to provide incentives to enterprises and institutions in their territory to transfer technology to LDCs. However, the provision falls short of requiring the actual transfer of technology to LDCs, which has generally rendered this provision ineffective. Proactive engagement from the African Group could change this. Together with other LDCs, African LDCs could identify a list of technologies that would be critical to enable smart agricultural production (Aggad et al. 2023). Given that discussion on transferring green technologies falls within the mandates of both the Committee on Agriculture and the TRIPS Council, it would be important to involve both bodies in these discussions (World Trade Organization 2023e).

9.10 Addressing agriculture and the environment at the WTO

As discussed in earlier chapters, food systems both contribute to environmental challenges, including climate change and biodiversity loss, and are impacted by them (FAO, UNDP and UNEP 2021). At the WTO, aligning trade and the environment has received increased attention. According to Director General Ngozi Okonjo-Iweala, 'trade and the WTO, are part of the solution to climate change and environmental degradation' (World Trade Organization 2021). Recent initiatives that seek to put environmental consideration at the heart of trade discussions include the TESSD, the Informal Dialogue on Plastics Pollution and Sustainable Plastics Trade and the Fossil Fuel Subsidy Reform. The FSA, which as we have seen aims to curb harmful fishery subsidies, is another important component of the WTO's sustainable trade initiatives. While negotiations on environmental goods and services have not progressed, there remains widespread interest in taking them forward.

Under the TESSD, participating WTO Members have established informal working groups focusing on trade-related climate measures, the circular economy, subsidies, and environmental goods and services. Each working group aims to advance ways in which trade can be used as a lever to address the respective climate and environmental challenges. From an agriculture and African food security perspective, the initiatives that are most relevant include the subsidy reform discussions in the Working Group on Subsidies, and the tariff and related discussions in the Working Group on Environmental Goods and Services.

In 2014, 46 WTO Members (not including any African countries)[13] launched plurilateral negotiations for the establishment of an Environmental Goods Agreement (EGA) to promote trade in key environmental products such as wind turbines and solar cells. It sought to do so by reducing or eliminating tariff and non-tariff measures on environmental goods. The negotiations reached a dead-end in 2016, in part because of disagreement on what constitutes an environmental good. The Working Group on Environmental Goods and Services in TESSD is seeking to revitalise the EGA negotiations and expand its scope to include services. These negotiations could positively affect food security in Africa if goods and services relevant to developing more resilient and high-yielding food products could be included within its scope. Lowering tariffs on various sustainable agricultural technologies such as early-warning systems, storage, and supply-chain infrastructure could help to increase food production. Participating African WTO Members could contribute to this discussion by identifying the types of goods and services that are critical from an agricultural production and food security perspective.

Similarly, issues related to post-harvest waste, addressed in the Working Group on the Circular Economy would be important from an African food security perspective. As much as 37 per cent of all food produced in Africa is lost between production and consumption (FAO 2011). Lowering barriers to trade in goods and services would enable the uptake of more circular agricultural production systems.

The Working Group on Subsidies focuses on addressing environmentally harmful subsidies, including agricultural subsidies. Subsidies linked to the production of a specific agricultural commodity, typically the staple crops, beef and poultry, generate environmentally harmful outcomes through overuse of agrochemicals and natural resources and contribute to nitrogen pollution and GHG emissions (Calvo 2022b). Some WTO Members are advocating the repurposing of agricultural subsidies towards addressing environmental concerns (Calvo 2022b; World Trade Organization 2023f). This would require diverting funding from agricultural subsidies with harmful environment effects to agricultural activities that promote better environmental outcomes (e.g. sustainable land management practices, or compensating farmers for ecosystem services like averting water runoff and soil erosion or offsetting GHG emissions). In the context of the AoA, this would mean that trade-distorting subsidies that would otherwise have been listed in the Amber Box will now come under the Green Box subsidies.

From an African food security perspective, repurposing subsidies is another aspect of the conundrum in the nexus between trade rules and food security objectives. On the one hand, research has shown that subsidies coupled to specific commodities result in higher levels of agricultural production (Calvo 2022b). On the other hand, repurposing agricultural support to achieve better environmental outcomes will likely reduce the volume of food that is produced globally with implications for food availability. Moreover, the anticipated box shifting that will happen because of the repurposing of domestic support – from Amber to Green – would essentially mean that WTO Members have no limits on domestic support that is linked to climate sustainability since there are no caps on support that can be provided under the Green Box. This will deepen the existing asymmetries between WTO Members.

As trade-offs can be made on a case-by-case basis, it would be important for the African Group to take a seat at the table and influence these discussions. As of October 2023, however, the 75 Members participating in the TESSD include only four African countries: Cabo Verde, Chad, Senegal and Gambia. While the Members of the African Group have expressed reservations about engaging in environmental discussions at the WTO, a recent submission by the African Group, 'Principles Guiding the Development and Implementation of Trade-Related Environmental Measures', suggests an increased openness to recognise the WTO as an institution to discuss trade and environment issues. (Lamy et al. 2023).

Summary

This chapter has examined the conundrum between WTO rules and global food security from an African perspective. The conundrum is manifested in the contradictory implications of WTO rules. Policymakers and negotiators must be aware of the many trade-offs that the conundrum implies. Five key trade-offs stand out.

Global vs. African agricultural production: Agricultural subsidies, enabled by FTBAMS allowances and *de minimis* thresholds, increase the global availability of food supply but also suppress commodity prices. This is beneficial from a global food security perspective since it means that more people have access to food at affordable prices. It also enables African NFIDCs to access the food they need. At the same time, large market-distortive, price-suppressing subsidies harm African agricultural production as farmers are not able to compete with the lower prices in the absence of government subsidies. For the African countries to become more food-secure, more food needs to be produced at home. Without reducing the FTBAMS of large agricultural producers, this will be practically impossible, especially for staples like rice, wheat, maize, meat and poultry.

National vs. global food security: Imposing export restrictions during periods of food shortages could be beneficial at a national level – at least in the short run – as it makes more food available at the national level. When many countries adopt the same measure, as was the case during the Covid-19 pandemic and following Russia's invasion of Ukraine, the result is an increase in global food prices and disrupts supply chains with catastrophic consequences for NFIDCs.

Export restrictions and price hikes vs. import surges and price suppression: Neither export restrictions with associated price hikes nor import surges associated with price suppression are desirable. Export restrictions and price hikes could result in severe food shortages for NFIDCs, whereas import surges and price suppression disincentivises agricultural production. While this trade-off between consumers and producers is not exclusive to agricultural trade, the tension is more pronounced with respect to food, especially staples.

Policy space for African countries vs. policy space for large developing countries: As long as African countries, most of which are NFIDCs, are grouped with large emerging markets like India and China as developing countries, successful negotiation by the African Group for carve-outs or exemptions will be difficult. These tensions have come to the surface especially in the context of public stockholding, *de minimis* allowances and the Development Box.

Food security vs. the environment: Another challenge as we saw is the tension between incentivising production and reducing the harmful impacts of agricultural subsidies through repurposing. While repurposing agricultural subsidies would be desirable from an environmental perspective, it could reduce the global amount of food produced, with potentially negative effects for African (and global) food security.

These trade-offs must be carefully navigated as the African Group seeks to make sure that WTO rules serve its food security objectives.

Agricultural negotiations remain contentious at the WTO, with limited progress in addressing imbalances and asymmetries. However, it is in the interest of the African group to work towards revitalising agricultural trade reform and also to call for new research that can offer fresh insights. This chapter has unpacked the issues, implications and conundrums with respect to domestic support, public stockholding, SSM, export restrictions, SPS, fishery subsidies, technology transfer, and trade and environment. Trade-offs are inevitable as some reforms might be desirable from an African agricultural production perspective but not from a consumption perspective. Others would not only secure policy space and flexibilities for African Members but would simultaneously provide benefits for developing countries with large agricultural production volumes, like China and India – with potentially negative implications for African agricultural producers. As we saw in Chapter 7, African countries have become export markets for these countries.

One critical aspect that stands out throughout this chapter is the importance of limiting benefits, such as those set out in the Development Box, to a subset of developing countries and conversely to apply proposed limits, for example caps on product-specific domestic support, or on support provided under paragraphs 5–12 of the Green Box, only to large agricultural producers. Whether to limit the Development Box to NFIDCs and LDCs, or to WTO Members that produce less than X per cent of global agricultural value, has to be negotiated and reflected upon. Upper thresholds that must be reached for specific restrictions to be applied must also be further explored.

African Members would also be advised to be pragmatic in agricultural negotiations, i.e. adopting an approach that focuses on results over principles, technical analysis over ideological positioning (van der Ven and Luke 2023). Some of the current positions advocated by the African Group on food security might not yield many benefits for Africa, or might not sufficiently strengthen African countries' food security situation. For example, in the context of SSM price suppression discussions, African countries should not overlook the flexibility they have in their tariffs schedules and use this in the event of price suppression. Given the highly politicised nature of the negotiations, it would be important for African countries to focus on areas that will have the most important impact from a food security and broader development perspective.

Another important observation that can be drawn from this chapter is the matter of implementation. While African WTO Members focus on agricultural negotiations, many of them have not used the Development Box and are not providing domestic support up to their allowed *de minimis* levels. This suggests that the problem is not necessarily a lack of policy space but also national policies and priorities, as discussed in Chapter 4.

Food security, agriculture and the environment are discussed in different fora and committees at the WTO including the Committee on Agriculture, the Work Programme on Food Security for LDCs and NFIDCs, the TRIPS Council, the Committee on Trade and Development, the SPS Committee and the Informal Working Group on Subsidies under the TESSD. It would be important for WTO Members to avoid discussing different aspects of food security in silos. This calls for enhanced cooperation between these relevant bodies, to streamline the discussions. With very small delegations in Geneva, African countries will surely benefit from a rationalisation of the food security agenda at the WTO. This is also necessary to understand the trade-offs better and to make sure adequate approaches are adopted and effective solutions are reached. This could be done through a Global Triangle Forum at the WTO, focused on matters at the intersection of trade, environment and development (Calvo 2022b).

While WTO rules can address market distortions and alleviate supply-chain shocks, trade remains only one among many considerations that impact Africa's food security. It has been shown throughout this book that low levels of agricultural output are a function of many factors, from climate change to technology applications, from finance and investment to productivity and production at scale. As shown in this chapter, the WTO legal framework is itself constrained by conundrums that cannot be easily resolved. Ultimately, African countries' policy choices and implementation processes at home are also critical.

Notes

[1] Article 1(a) of the AoA defines aggregate measurement of support (AMS) as follows: the annual level of support, expressed in monetary terms, provided for an agricultural product in favour of the producers of the basic agricultural product or non-product-specific support provided in favour of agricultural producers in general.

[2] Uniquely, China has a *de minimis* entitlement of 8.5 per cent.

[3] See Agreement on Agriculture, 1994, Annex 2.

[4] See Ambaw et al. (2021).

[5] This was negotiated as part of China's WTO Accession Protocol.

[6] One exception is Zambia, which notified Article 6.2. spending at 8 per cent of the value of production in 2000.

[7] The countries that have requested consultations include the US, the EU, Australia, Canada, Japan, Brazil, Paraguay, Uruguay and Thailand. See Mishra (2022).

[8] Also see Committee on Agriculture (2015).

[9] As mentioned above, tariffication refers to the process of replacing agriculture-specific non-tariff measures with a tariff that affords an equivalent level of protection.

[10] See World Trade Organization (2002).

[11] All these countries have developing country status in the WTO, which allows Members to announce whether they are 'developed' or 'developing' countries.

[12] It must be noted that even when WFP is delivering food assistance, this assistance often gets abused. For example, in some African countries government officials and/or the private sector have sold to make a profit the WFP's delivered food, which was meant to be provided free of charge to the hungry. See e.g. World Food Programme (2023); Bailey (n.d.).

[13] If the individual members states of the European Union are also counted. The WTO Members that were part of the initiative were: Australia; Canada; China; Costa Rica; the EU; Hong Kong, China; Iceland; Israel; Japan; Korea; New Zealand; Norway; Singapore; Switzerland; Liechtenstein; Chinese Taipei; Turkey; and the US.

References

African Development Bank (2022) 'The Future of Marine Fisheries in the African Blue Economy'.
https://www.afdb.org/en/documents/future-marine-fisheries-african-blue-economy

African Group (2023) 'Domestic Support JOB/AG/242 [Restricted]', 9 June.

African Group; the ACP; and G33 (2022) 'Public Stockholding for Food Security Purposes, Proposal by the African Group, the ACP and G33', WT/MIN(22)/W/4, 6 June.

Agreement on Agriculture, 1994.

Aggad, Faten et al. (2023) 'Implications for African Countries of a Carbon Border Adjustment Mechanism in the EU', African Climate Foundation and The London School of Economics and Political Science. https://perma.cc/TVQ2-NY4P

Ambaw, Dessie et al. (2021) 'Strengthening African Agricultural Trade: The Case for Domestic Support Entitlement Reforms', Institute for International Trade Working Paper 07, August.

Anderson, Kym et al. (2021) 'Impacts of Agricultural Domestic Supports on Developing Economies', New Zealand Institute for Business Research, November. https://perma.cc/8355-8G2S

Avesani, Cosimo (2023) 'Overview of Public Food Stockholding Programmes: Policies and Practices', PSH Rules to Promote Food Security of all WTO Members, 28 September. https://perma.cc/U5AR-LR24

Bailey, Sarah (n.d.) 'Somalia Food Aid Diversion', ODI: Think change. https://odi.org/en/insights/somalia-food-aid-diversion

Benin on behalf of the African Group (2019) 'African Group Elements on Agriculture: For Meaningful Development Outcomes at the Twelfth Ministerial Conference', JOB/AG/173.

Calvo, Facundo (2022a) 'How Can the WTO Contribute to Global Food Security?', IISD SDG Knowledge Hub. https://perma.cc/79S7-9SS8

Calvo, Facundo (2022b) World Trade Organization Talks on Agricultural Subsidies Should Consider Trade-Offs Among Trade, Food Security, and the Environment, IISD International Institute for Sustainable Development. https://perma.cc/W35U-GG4F

Calvo, Facundo (2023) 'WTO Members Should Avoid Export Restrictions on Fertilizers to Deliver on Food Security', IISD International Institute for Sustainable Development. https://perma.cc/9ELA-CD8G

Committee on Agriculture (2021a) *Submission by Brazil: Article 6.2 of the Agreement on Agriculture (AoA) in perspective, JOB/AG/195.*

Committee on Agriculture (2021b) G/AG/N/IND/25, 8 April.

Committee on Agriculture (2022a) 'Committee on Agriculture', G/AG/N/IND/27, 1 April.

Committee on Agriculture (2022b) 'Questionnaire on LDC and NFIDC Members' Utilization of WTO Flexibilities (Work Programme – Paragraph 8 of MC-12 Declaration On Food Insecurity)', G/AG/GEN/214, 8 December.

Committee on Agriculture (2022c) 'Treatment of LDCs and NFIDCs under the WTO Agriculture Rules, Note by the Secretariat', G/AG/W/227, 29 November.

Committee on Agriculture (2023a) G/AG/N/IND/29.

Committee on Agriculture (2023b) 'Special Session (2023) Special Safeguard Mechanism for Developing Country Members: Communication from the African Group', JOB/AG/205/Rev.1 [Restricted].

Committee on Agriculture (2023c) 'Why MC13 Needs to Address Export Restrictions on Agricultural Products', JOB/AG/244.

Committee on Agriculture (2023d) 'Summary of Responses to the Questionnaire on LDC and NFIDC Member's Utilization of WTO Flexibilities Note by the Secretariat', G/AG/W/23. https://perma.cc/2YMF-PSAY

Das, Abhijit et al. (2020) 'Special Safeguard Mechanism for Agriculture: Implications for Developing Members at the WTO', Centre for WTO Studies, IIFT New Delhi, Working Paper No. CWS/WP/200/59. https://perma.cc/CA87-4CV3

Decision on Measures Concerning the Possible Negative Effects of the Reform Programme on Least-Developed and Net Food-Importing Developing Countries, 1994.

Directorate-General for Agriculture and Rural Development (n.d.) 'The World Trade Organization and EU Agriculture', European Commission, Agriculture and rural development. https://perma.cc/B8UC-VUM7

Espitia, Alvaro; Rocha, Nadia; and Ruta, Michele (2022) 'How Export Restrictions Are Impacting Global Food Prices', *World Bank Blogs*, July. https://perma.cc/UQK5-77CH

FAO (ed.) (2011) 'Global Food Losses and Food Waste: Extent, Causes and Prevention. International Congress Save Food!', Rome: Food and Agriculture Organization of the United Nations.

FAO; UNDP; and UNEP (2021) 'A Multi-Billion-Dollar Opportunity – Repurposing Agricultural Support to Transform Food Systems', Rome: FAO. https://doi.org/10.4060/cb6562en

Fevrier, Stephen; and Dugal, Manleen (2017) 'The WTO's Role in Fisheries Subsidies and Its Implications for Africa', *tralac*, 6 January. https://www.tralac.org/news/article/11038-the-wto-s-role-in-fisheries-subsidies-and-its-implications-for-africa.html

Food and Agriculture Organization of the United Nations; World Trade Organization; and World Bank (2023) *Rising Global Food Insecurity: Assessing Policy Responses*, April. https://perma.cc/FCW4-2VEN

Food and Land Use Coalition (2019) 'Growing Better: Ten Critical Transitions to Transform Food and Land Use'. https://perma.cc/Y4HQ-XYMF

G-33 (2017) 'The Short-Term Price Volatility in Agriculture: Need for Stability for Small-Scale Farmers in Developing Country Members', Submission TN/AG/GEN/45.

General Agreement on Tariffs and Trade (GATT), 1994.

General Council, Committee on Agriculture and Committee on
  Trade and Development (2023) 'The Role of Transfer of Technology
  in Resilience Building: Agriculture Communication from the
  African Group (Angola; Benin; Botswana; Burkina Faso; Burundi;
  Cabo Verde; Cameroon; Central African Republic; Chad; Congo;
  Côte D'ivoire; Democratic Republic of Congo; Djibouti; Egypt;
  Eswatini; Gabon; The Gambia; Ghana; Guinea; Guinea-Bissau;
  Kenya; Lesotho; Liberia; Madagascar; Malawi; Mali; Mauritania;
  Mauritius; Morocco; Mozambique; Namibia; Niger; Nigeria; Rwanda;
  Senegal; Seychelles; Sierra Leone; South Africa; Tanzania; Togo;
  Tunisia; Uganda; Zambia And Zimbabwe)', World Trade Organization.
  https://perma.cc/2G8Q-YU3K

Giordani, Paolo E.; Rocha, Nadia; and Ruta, Michele (2012) 'Food Prices
  and the Multiplier Effect of Export Policy', World Trade Organization
  Economic Research and Statistics Division Staff Working Paper ERSD-
  2012-08, April. https://perma.cc/W3HA-V2EG

Glauber, Joseph et al. (2022) 'MC12: How to Make the WTO Relevant in the
  Middle of a Food Price Crisis', IFPRI. https://perma.cc/J4CQ-U6MX

Gro Intelligence (n.d.) 'Tracking Food Security Across Africa'.
  Previously available at:
  https://community.gro-intelligence.com/food-security-tracker-africa

International Monetary Fund (2022) 'Climate Change and Chronic
  Food Insecurity in Sub-Saharan Africa', IMF Departmental Paper
  DP/2022/016, September.

Jaffee, Steven et al. (2019) 'The Safe Food Imperative: Accelerating Progress
  in Low- and Middle-Income Countries', Agriculture and Food Series.
  Washington, DC: World Bank. https://perma.cc/CU3Y-QN4S

Lamy, Pascal et al. (2023) 'The Case for a Global Triangle Forum at the
  WTO'. https://perma.cc/ARD4-99TD

Lau, Christian; and van der Ven, Colette (2017) 'The WTO Agreement on
  Agriculture', in *ICC Business Guide to Trade and Investment*. Paris: International Chamber of Commerce.

Matthews, Alan (2014) 'Food Security and WTO Domestic Support Disciplines Post-Bali', ICTSD Programme on Agricultural Trade and Sustainable Development, Issue Paper no. 53, Geneva: International Centre for
  Trade and Sustainable Development.

Ministerial Declaration on the Emergency Response to Food Insecurity
  (WT/MIN (22)/28), 2022.

Mishra, Asit Ranjan (2022) 'India Opposes Group Consultation on Food
  Subsidies Programme at the WTO', *Business Standard*, 14 October.

Reuters (2022) 'WTO Nears Food Pledges; India, Egypt, Sri Lanka Hold Out', *Reuters*, 14 June. https://www.reuters.com/markets/commodities/wto-nears-food-pledges -india-egypt-sri-lanka-hold-out-2022-06-14/

Sinha, Tanvi; and Glauber, Joseph (2021) 'MC12: An Opportunity to Find an Enduring Solution on Public Stockholding'. https://perma.cc/L6C4-DZHG

Tenth WTO Ministerial Conference, Nairobi, 2015 (2024) 'Export Competition Ministerial Decision of 19 December 2015 : WT/MIN(15)/45 — WT/L/980', World Trade Organization. https://perma.cc/CKN6-B9S6

Trade and Markets Division (2009) 'Part 1 What Happened to World Food Prices and Why?', in *The State of Agricultural Commodity Markets 2009*, Rome: FAO, pp.8–29. https://perma.cc/J5SB-CHU8

van der Ven, Colette; and Luke, David (2023) 'Africa in the World Trade Organization', in Luke, David (ed.) *How Africa Trades*, London: LSE Press. https://doi.org/10.31389/lsepress.hat

Walker, Timothy; Reva, Denys and Willima, David. (2022) 'Does Africa Gain from the WTO's Landmark Fishing Subsidies Deal?', Institute for Security Studies, 8 September. https://perma.cc/G6XW-X53D

WHO (2015) 'WHO Estimates of the Global Burden of Foodborne Diseases'. https://apps.who.int/iris/bitstream/handle/10665/199350/9789241565165 _eng.pdf

World Food Programme (2023) 'Widespread Food Diversion Impacts WFP Food Distributions across Ethiopia', WFP World Food Programme. https://perma.cc/EZ7B-XHVM

World Bank; and United Nations Department of Economic and Social Affairs (2017) *The Potential of the Blue Economy: Increasing Long-term Benefits of the Sustainable Use of Marine Resources for Small Island Developing States and Coastal Least Developed Countries*, Washington DC: World Bank.

World Trade Organization (2002) 'WTO Secretariat background paper "Special Agricultural Safeguard"', G/AG/NG/S/9/Rev.1.

World Trade Organization (2013) 'Ministerial Decision of 7 December 2013, Public Stockholding for Food Security Purposes', WT/MIN(13)/38, WT/L/913.

World Trade Organization (2014) 'The Bali Decision on Stockholding for Food Security in Developing Countries Updated 27 November 2014', World Trade Organization. https://perma.cc/J56A-SSAS

World Trade Organization (2015) 'Tenth WTO Ministerial Conference, Nairobi, 2015, Briefing Note: Agriculture Issues'. https://perma.cc/26CE-MERB

World Health Organization (2015) 'WHO Estimates of the Global Burden of Foodborne Diseases: Foodborne Disease Burden Epidemiology Reference Group 2007–2015', Geneva: World Health Organization. https://iris.who.int/handle/10665/199350

World Trade Organization (2020) 'Committee on Agriculture, G/AG/N/IND/18', 31 March.

World Trade Organization (2021) 'Trade and Environment. New Initiatives Seeks to Put Environment at Heart of Trade Discussions', World Trade Organization. https://perma.cc/3VJL-QJZL

World Trade Organization (2022a) 'African Ministers of Trade Declaration on WTO Issues', WT/MIN (22)/10.

World Trade Organization (2022b) 'Ministerial Decision on World Food Programme Food Purchases Exemption from Export Prohibitions or Restrictions', WT/MIN (22)/29, 22 June.

World Trade Organization (2022c) 'Implementing the WTO Agreement on Fisheries Subsidies: Challenges and Opportunities for Developing and Least-Developed Country Members'.

World Trade Organization (2022d) 'Sanitary and Phytosanitary Declaration for the Twelfth Ministerial Conference: Responding to Modern SPS Challenges', WT/MIN (22)/27, 2022. https://perma.cc/K296-ZZ6S

World Trade Organization (2023a) 'Agreement on Agriculture (AOA): The Amber and Blue Box, Product-Specific Concentrations of Support Submission by the Cairns Group', World Trade Organization. https://perma.cc/JP2Q-VDT7

World Trade Organization (2023b) 'DDG Ellard Highlights Importance of WTO Work on Fisheries Subsidies for Africa', World Trade Organization. https://perma.cc/5MFG-PNJM

World Trade Organization (2023c) 'WTO Workshop on Fisheries Subsidies for English-Speaking Africa', World Trade Organization. https://perma.cc/C645-QZ5M

World Trade Organization (2023d) 'WTO Members Advance Fisheries Subsidies Negotiations at Second Fish Week', World Trade Organization. https://perma.cc/8ZPP-QRTD

World Trade Organization (2023e) 'Working Group on Food Security Discusses Building Resilience, Financing Challenges', World Trade Organization. https://perma.cc/3ZJA-RGBG

World Trade Organization (2023f) 'DDG Paugam Urges Ministers to Harness WTO Potential for Advancing Agricultural Innovation', World Trade Organization. https://perma.cc/KWF8-7GNT

World Trade Organization (2024a) 'Food Security', World Trade Organization. https://perma.cc/6A9W-KRSF

World Trade Organization (2024b) 'Agriculture: Explanation. Market Access', World Trade Organization. https://perma.cc/N5TY-R2VL

World Trade Organization (2024c) 'Agriculture: Explanation Domestic Support', World Trade Organization. https://perma.cc/UN23-DX55

World Trade Organization (2024d) 'Agriculture: Explanation. Export/Competition/Subsidies', World Trade Organization. https://perma.cc/QHE3-JRRS

World Trade Organization (2024e) 'Domestic Support in Agriculture: The Boxes', World Trade Organization. https://perma.cc/AU9Z-AV6K

World Trade Organization (2024f) 'Agriculture Negotiations', World Trade Organization. https://perma.cc/V5AN-MLDA

WTO (2013) 'Decision on General Services, Ministerial Decision on 7 December 2013', WT/MIN(13)/37, 11 December.

WTO General Council (2014) 'Public Stockholding for Food Security Purposes Decision of 27 November 2014', World Trade Organization.

10. Conclusion: trade, food security and climate risks

David Luke

This is a book that is neither wholly about agricultural policy nor wholly about trade policy nor wholly about climate policy. The research team that came together under the Africa Trade Policy Programme at the London School of Economics Firoz Lalji Institute for Africa sought to understand and explain why Africa struggles with food availability and stability, which are the essential pillars of food security. We took the intersection between trade, agriculture and climate policies as the point of entry for our enquiry into why 280 million Africans, a fifth of the continent's population, live with malnutrition and 340 million Africans, a quarter of the population, face hunger. We applied analytical tools and data to these policy areas, which enabled us to alight on some insights on why food deprivation on this scale persists in Africa.

We established that the continent's status as a net food importer has stabilised over the last decade in absolute terms and has not worsened despite rapid population growth and rising per capita incomes during much of this time. In 2021, Africa as a whole recorded an annual net trade deficit of $34 billion in the food and agricultural sector but below the peak of $47 billion reached in 2011 (in nominal prices, which suggests underappreciation of how much more significant the deficit was in 2011). If we read gains in productivity and output into these figures as fairly good news for the continent as a whole, the sobering reality is that food insecurity remains widespread in 42 of the 54 African countries, as the headline numbers on food deprivation attest.

We traced the root of Africa's food security challenges to an economic structure that is based on mainly unprocessed, primary products being exported in return for imports of final consumption goods. Lack of economic diversification from primary commodities has been described as 'the heart of the matter' for African development (Mangeni and Mold 2023). The implication

---

How to cite this book chapter:

Luke, David (2025) 'Conclusion: trade, food security and climate risks', in: Luke, David (ed) *How Africa Eats: Trade, Food Security and Climate Risks*, London: LSE Press, pp. 243–255. https://doi.org/10.31389/lsepress.hae.j License: CC-BY-NC 4.0

is that Africa's commodity trade supports value addition, economic growth and jobs elsewhere in the world while reinforcing high rates of poverty at home. Since poverty, unemployment and food insecurity are interrelated, it is an economic model that is inherently vulnerable to food security risks, which is intensified when terms of trade shift or shocks emerge, as the recent Covid-19 pandemic and post-Covid-19 food price inflation attest. The Economic Commission for Africa of the United Nations estimates that 50 million more Africans were in poverty in 2023 than in 2019, an increase of 28 per cent (United Nations Economic Commission for Africa 2024). Africa's population grew by an estimated 10 per cent over the same period, so this largely driven by a rise in the poverty rate (not only population growth).[1] Overall, 476 million Africans – about a third of the population – were in poverty in 2023. As poverty rates have declined elsewhere in the world, 60 per cent of the world's extreme poor now live in Africa. Food deprivation is inherently a symptom of poverty. An assessment of the economic growth requirements for achieving the sustainable development goal of halving poverty by 2030 in African countries concludes that, on average, African countries will not only need to grow by 6 per cent or more every year up to 2030 but also need to ensure that the benefits of growth are widely shared through social protection policy measures that take prevailing inequalities into account (Fofana, Chitiga-Mabugu and Mabugu 2023).[2]

We recognised climate change as a factor that is making an already-trying agriculture and food security situation even more difficult. Agricultural emissions are also part of the problem. This requires greater scrutiny of the sustainability of production systems. We outlined an approach for thinking about the interaction between trade, food security and climate risks and identified the varying effects of climate change particularly on the production of the eight food products most widely consumed in Africa, which we referred to as Africa's basic foods.

The eight basic products of yam, cassava, maize, rice, wheat, meat, poultry and fish contribute significantly to daily calorific intake across the continent. We established that yields in almost all of these products generally trailed global productivity and output, despite nominal growth in production. Yams are the main exception and cassava a partial exception.

Agricultural policies are important determinants of food security outcomes. Finance, investment, foreign aid, institutions, actors and capacities interact with policies in playing a key role in resource allocation along the food value chain, from production to consumption, from supply to demand. We unpacked the chokepoints in policy implementation, resources, capacities, climate and sustainability risks that hold Africa back from becoming an agricultural powerhouse despite having 60 per cent of the world's arable land area.

We probed the observable patterns of food trade within the continent or at the intra-African level and with foreign partners. A major consequence of underperformance in Africa's food production is that intra-African trade in food products remains relatively small, although this trade has grown in value in real terms over the last 10 years. This is in line with our finding that the continent's status as a net food importer has stabilised and not worsened. Facilitating greater intra-African trade could boost the continent's food production if it allows African producers to gain a greater share in the African market. Small-scale informal cross-border trade is ubiquitous and reflects the dominance of the smallholder farming model. By creating a detailed partial equilibrium model to simulate the expected impact of the African Continental Free Trade Area (AfCFTA), a major initiative to liberalise trade across the continent, we found that the impact of the AfCFTA on intra-African trade will be relatively modest. That is because much of that trade is already liberalised through pre-existing regional trade agreements, such as those of the EAC, COMESA, SADC and ECOWAS. It is through these regional arrangements that most of Africa's *current* intra-African trade in the agriculture sector flows.

This finding on the limited gains to be expected from the AfCFTA prompted us to examine the provisions of the AfCFTA Agreement and Protocols on non-tariff barriers. We found that the AfCFTA legal instruments embody best practices for harmonisation of food safety standards, technical regulations and regulatory compliance. If these provisions, along with other regulatory measures on services, investment, digital trade, competition policy and intellectual property rights, are implemented effectively, they could boost intra-African value chains in agriculture and agribusinesses, enhance efficiency and lower prices.

Concerning trade with foreign partners, we deconstructed the food trade deficit to reveal that the geography of these trade relationships is changing, with increasing food trade flows between African and emerging partners in the Global South. The changing geography is also reflected in trade in agricultural inputs as the European Union (EU) and United States (US) lose trade shares to China and India.

Finally, we turned our attention to the World Trade Organization (WTO), the world's trade regulator, where 42 African countries are part of the net food-importing developing countries group that coordinates efforts to keep international food markets open, monitors food aid flows and constitutes an important stakeholder group in negotiations to reform global rules on agricultural trade. We examined the contradictory role that food security plays in international trade. Agricultural subsidies in countries that can afford them incentivise food production not only for domestic consumption but also for trading in open markets and for food aid donations. While overproduction contributes to global food availability, it disincentivises production in poorer and net food-importing countries. The trade-offs in multilateral agricultural reforms need to be better understood by African negotiators as some reforms might be desirable from an African agricultural production perspective but not necessarily from a consumption perspective.

This leads us to the following conclusions, which are further elaborated on below:

1. Climate change poses multiple risks to food security.
2. Food production is responding to population pressure and expanding demand but not sufficiently to close the food trade deficit.
3. Productivity trails global levels for the vast majority of the basic foods that Africans commonly consume.
4. Gaps in agricultural policy implementation, finance, institutions and capacities enhance food security vulnerabilities.
5. Food dominates intra-African trade, which remains small despite growth in value in real terms over the past decade.
6. The impact of AfCFTA tariff liberalisation on food trade flows will be modest.
7. AfCFTA provisions on non-tariff barriers will have greater impact on food trade flows.
8. Beyond the AfCFTA, the reality is that most African countries are net food importers and increasingly source food imports from countries of the Global South.
9. Agricultural negotiations remain contentious at the WTO, with limited progress in addressing imbalances and asymmetries, but trade-offs implied in agricultural reforms need to be better understood.

10.1 Climate change poses multiple risks to food security

The risks of a changing climate on agricultural production cannot be over-estimated. The varying effects of climate change on food production – in particular, rising temperatures, extreme weather variations and the frequency of adverse supply shocks – were outlined in a model we presented in Chapter 2 for thinking about the interaction between trade, food security and climate risks. In Chapter 3 we applied the model to identify risks such as water stress; shortened crop growing seasons; shrinking acreage of arable land; higher incidence of crop pests; inundation of cropland and erosion; and flood-induced damage to agriculture-related infrastructure. We noted that the indirect effects are equally impactful and include reduced labour productivity of farm workers, whether due to harsh climatic conditions or illness as vector-borne diseases proliferate and disincentive effects leading some farmers to abandon their farms altogether. Yields are projected to fall for most staple crops and from livestock across most of Africa, including important sources of food security such as wheat, maize, rice and meat.

At the same time, we recognised that agricultural activities such as enteric fermentation of ruminant livestock and irrigated rice farming practices are significant contributors to methane emissions and other greenhouse gases (GHGs). Land itself is both a source and a sink of carbon emissions.

The nationally determined contributions (NDCs) of African countries are replete with a variety of adaptation and mitigation measures. But NDCs remain a wish list in the absence of adequate financing to implement them. NDCs, which are set for multiple years in advance, also need to be adaptive to fast-changing knowledge and technologies that can be applied to adaptation and mitigation strategies.

Trade can help to reduce the impact of production shocks, including those affecting critical food security crops. As we observed, adverse supply shocks in certain places can be met by supply surpluses in other places through trade. But availability is only part of any solution as food security also requires purchasing ability to ensure access. Endemic poverty in Africa is among the factors that limits ability to access food.

We noted that trade in agricultural intermediates and inputs, as well as agricultural services and knowledge, can play an important role in agricultural adaptation to climate change. These help farmers utilise new seed varieties, agricultural machinery, fertilisers and agricultural extension services to address changing climate challenges. Trade reforms to reduce tariff and non-tariff barriers can help to facilitate access to these inputs.

10.2 Food production is responding to population pressure – but not enough

We unpacked the composition of the continent's persistent trade deficit in agriculture in Chapter 2 by reviewing the agriculture sector as a whole, including trade in basic foods like grains, tubers, meat and poultry, agricultural commodities such as cocoa, tobacco, coffee, tea and spices and trade in inputs like fertilisers and pesticides and capital equipment like farm machinery. We noted that, in 2021, African countries had an annual net trade deficit of $49 billion in basic foods and $9 billion in agricultural capital, while returning a net surplus of $16 billion in exports of agricultural commodities and of $6 billion in exports of agricultural inputs. This deficit widened dramatically from the early 2000s to 2011 before stabilising in the last decade, with the $36 billion deficit for the sector as a whole in 2021 being about a quarter smaller than it was in 2011.

As these are nominal figures, they underappreciate how much more significant the deficit was in 2011 and the steady if also gradual trend in closing the gap. However, the median African country spends a quarter of the revenue it earns through exports on food imports, while 16 countries spend more than 40 per cent on food imports. These countries risk serious food insecurities if adverse terms of trade shocks arise or if world food prices rise.

We therefore concluded that the agricultural trade deficit is driven by food imports along with production and export underperformance. Food imports include low-unit-value foods, such as cereals like wheat, rice and maize but also some higher-unit-value foods like fish, dairy and poultry. Demand for the former grows fastest with population growth, the latter with rising per capita incomes. But it is also important to note that the deficit in Africa's food and agriculture sector has been reasonably stable over the last decade, despite both population growth and rising per capita incomes during this time.

10.3 Productivity trails global levels for most of Africa's basic foods

We inquired more fully in Chapter 3 into the role of yams, cassava, maize, rice, wheat, meat, poultry and fish as products or basic foods with high rates of per capita consumption in Africa. We distinguished between regional variations in production and consumption. We established that yields in the production of almost all of these basic foods generally trailed global productivity and output, although there has been an increase in production. With yams and cassava as the exceptions, the other foods are not produced at a scale and at levels of productivity that is sufficient to meet demand. While the comparative advantage of the African cassava- and yam-producing countries remains incessant, this is not the case for rice, wheat and maize, beef or poultry, which benefit from significant subsidies in richer countries, with trade-distorting effect. For maize specifically, Africa produces more than it consumes for food but the continent is a net importer. This is because the excess production over human consumption is not sufficient to meet demand for other uses of maize including feed for livestock and industrial processing and manufacturing.

We highlighted the challenges facing the fish sector, which include underinvestment in the management of fish stocks, the marine environment and freshwater habitats, illegal unregulated and unreported fishing by foreign boats and rising sea temperatures. As the sea temperatures rise, fish stocks migrate towards colder waters. This increases pressure on small-scale fishing communities to scale up operations by investing in equipment and vessels that can go out further into the sea.

10.4 Gaps in policy implementation, finance, institutions and capacities increase food security vulnerabilities

Having identified the climate risks, established what drives the agricultural trade deficit and considered productivity and output of the most widely consumed foods, we turned our attention in Chapter 4 to assess the policy and institutional issues that contribute to the vulnerabilities that were observed. On agricultural policy, we explained that African Union policy frameworks such as the Comprehensive African Agriculture Development Programme (CAADP) and the Malabo Declaration provide African countries with a blueprint for boosting agricultural development and trade, to achieve the much-vaunted green revolution. CAADP requires governments to allocate at least

10 per cent of public expenditure to agriculture and to aim for 6 per cent annual growth in the sector. Reviews, however, suggest that only one country – Rwanda – is on track to achieving the CAADP goals. Financial resources remain a major constraint. While there are many good examples of the impact of agricultural financing, there is scope for scaling up private investment, farmers' access to credit, foreign direct investment, foreign aid and, increasingly of importance, climate finance. Development partners provide relatively little assistance to agricultural development in Africa despite the clear understanding that this sector is critical for achieving international goals on poverty and hunger and overcoming gender inequalities. When it comes to food aid – a convenient channel for dumping by food surplus countries – we argued that this needs to be carefully managed in order not to disincentivise local production.

We saw that capacities vary among actors and institutions that mediate production, markets and value chains such as farmers, 'middlemen', cooperatives, commodity exchanges and agricultural marketing boards. We suggested that partnerships with multinational corporations that play a dominant role in global food supply chains can be beneficial where local interests are well safeguarded. With the bulk of African agriculture still in the hands of small-scale farmers, measures to boost production and productivity must necessarily focus on smallholders. The rise of contract farming and a class of medium-scale farmers are promising developments, especially since this class of farmers has stronger commercial ambitions than smallholders do. We argued that agricultural commercialisation is the most viable pathway for smallholders to increase their productivity, output, income and food security but there are huge challenges as regards imperfect or missing markets and institutions.

10.5 Food dominates intra-African trade

We probed in Chapter 5 the observable patterns of food trade within the continent or at the intra-African level. We noted that a major consequence of low productivity and output in agricultural production is that intra-African trade while dominated by trade in food products remains relatively small, although this trade has grown in value in real terms over the last 10 years. Cereals, tubers, vegetables and fruits, fish and fish preparations are the main food products that are traded. Food security in tubers has been achieved but African countries import almost twice as many cereals from the rest of the world as they do from each other. The smallholder farming model has given rise to small-scale informal cross-border trade that is widespread. Although difficult to quantify, estimates suggest that informal trade could be as high as 16 per cent of total formal intra-African trade and as much as 72 per cent of trade between neighbouring countries.

Each of the basic foods tends to be traded within Africa in its own way. The vast majority of trade in cassava takes place in Eastern Africa. Yams are mainly traded in Southern and Western Africa. Trade in maize is concentrated in Eastern and Southern Africa. Rice is traded mainly in Eastern, Southern, Central and Western Africa, with intra-African rice imports in Northern Africa being negligible. Intra-African wheat trade is evenly distributed among the regions, except in Central Africa, which accounts for only a small share of these imports. Trade in meat and poultry also occurs predominantly within and between Eastern and Southern Africa. Almost half of intra-African fish imports occur in Western Africa, with the rest of this trade being more evenly split between Eastern, Central Africa, Southern and Northern Africa.

10.6 The impact of African Continental Free Trade Area tariff liberalisation on food trade flows will be modest

We reported in Chapter 6 the result of our partial equilibrium model to simulate the expected impact of the AfCFTA. The AfCFTA initiative is driven by the recognition that a liberalised trade regime across the continent could generate further growth in intra-African trade, including informal trade formalisation, as tariffs not already covered by regional trade agreements and non-tariff barriers fall. However, we found that the impact of the AfCFTA on intra-African trade will be relatively modest since much of the trade is already liberalised through pre-existing regional trade agreements across the continent.

Moreover, the AfCFTA is even less likely to have an impact on trade in cereals such as wheat, maize and rice. These products already have, on average, low tariffs and are mostly served by more efficient or highly subsidised suppliers outside the continent. We found that where the AfCFTA will have an impact in the immediate term is in the downstream consumable food part of the value chain, and especially with higher-unit-value foods like fish and seafood, vegetables, cereal preparations, vegetable oils, fruits and dairy. There are also relatively sizeable opportunities for exports of sugar and coffee, within agricultural commodities. In the upstream part of the value chain, though the prospects for trade creation are smaller overall, there are important opportunities for exporters of agricultural machinery, fertilisers and pesticides. South Africa, for example, might begin to supply more of the continent's needs of agricultural machinery, while more fertilisers and pesticides could be supplied by North African countries like Morocco and Mauritania.

10.7 African Continental Free Trade Area provisions on non-tariff barriers will have greater impact

In view of the limited impact to be expected from AfCFTA tariff liberalisation on food trade, we examined in Chapter 7 the provisions of the AfCFTA

Agreement and Protocols on non-tariff barriers (NTBs). We found that the AfCFTA legal instruments embody best practices for the harmonisation of food safety standards, technical regulations and regulatory compliance. We noted that customs and trade facilitation provisions that aim to streamline border processes are critical for perishable goods. The creation of a web portal where traders and governments can submit complaints about partners' NTBs is an important initiative that could help to discipline unwarranted controls. The protocols on services, investment, competition policy, intellectual property rights, digital trade and small and medium-sized enterprises led by women and youth, if fully implemented, could boost intra-African value chains in agriculture and agribusinesses, enhance efficiency, and lower prices for consumers.

On services, we noted that some, including financial, logistics, information and communication technologies, insurance, distribution and transport services, are intrinsically linked to food systems through agricultural production, distribution and trade, and through these channels to food security. Logistics services such as transport, and information and communication technology, are critical to reduce costs and uncertainty in agricultural trade. With very low levels of financial inclusion among African farmers, increasing access to financial services through their liberalisation could enhance the uptake of financial services utilisation. Intra-African liberalisation of these and other services could attract investment and enhance competition with transformative impacts on agricultural production, value chains and food security.

The investment facilitation provisions of the investment protocol address constraints issues such as excessive bureaucracy, lack of transparency about investment-related information, corruption, and inadequate coordination among regulatory institutions. These are key issues that hinder intra-African investment flows that typically target the agricultural sector. However, the protocol requires investors and their investments to respect and protect the environment while carrying out their business activities. Among specific investor obligations are the right to a clean and sustainable environment, complying with the principles of prevention and precaution to anticipate significant harm to the environment, carrying out an environmental impact assessment, and mitigating and restoring any environmental harm that companies have caused.

With increasing economic concentration in the production and trading of agriculture and food products both globally and within the African continent, implementation of the Protocol on Competition Policy could play an important role in addressing anticompetitive behaviour in the food sector. It aims to discipline practices such as abuse of dominant positions in the market and mergers or acquisitions that restrict or prevent competition.

The Protocol on Intellectual Property Rights applies to all categories, including seed and plant varieties, geographical indications, genetic resources

and traditional knowledge. Putting such protections in place can incentivise investment in innovation in the development of new, higher-yielding, or
drought- and heat-tolerant plant varieties. But these protections need to be
balanced by adequate access and benefit sharing provisions, to ensure for
instance that farmers are not prevented from using new plant varieties. Safeguards are also needed to ensure that communities benefit from geographical
indications, genetic resources and traditional knowledge.

The Protocol on Women and Youth addresses the historical challenges
this category of farmers, entrepreneurs and business owners has faced, such
as access to trade finance, participation in trade policymaking, support to
enhance export capacity, and a range of trade facilitation measures that have
not been gender sensitive. These protocol's provisions can help to ensure
inclusivity in food production and trade.

The Protocol on Digital Trade is aimed at creating a digital enabling environment that can boost the uptake of digital technologies that are critical to
boosting agricultural yields and enhancing food preservation. This includes,
for instance, automated drip-irrigation technologies; digital technologies that
enable up-to-date tracking of produce that is being transported to markets
or access to information to optimise crop pests/disease mitigation strategies;
and mobile phone applications that set out early-warning systems regarding
weather events and access to real-time product prices.

10.8 Most African countries remain net food importers and increasingly source food from the Global South

Having established that implementation of the non-tariff provisions in the
AfCFTA could have a much greater impact on agricultural production and
agri-business, with significant benefits for food security, we returned in
Chapter 8 to analyse the current reality that 42 African countries are net
food importers. We examined how the food trade deficit breaks down with
Africa's trading partners or on a bilateral basis. The main insight that we
discovered is that the geography of these trading relationships is changing, with increasing food trade flows between African and countries in the
Global South. The changing geography is also reflected in trade in agricultural inputs such as machinery, seeds, fertilisers and herbicides, with
traditional partners such as the EU and the US losing trade shares to
emerging partners.

Brazil is the largest net food supplier to Africa, followed by the EU and
the US. The EU, however, remains Africa's most important market for both
food exports and imports. The EU and the US are also significant suppliers of
agricultural seeds, machinery and tractors to Africa. Among traditional partners, Russia and Ukraine are a major source of cereal exports to some African
countries. The Russia–Ukraine war that started in 2022 disrupted the flow of
these exports. But the concentration of Russia's and Ukraine's grain exports in

a few African countries limited a wider damaging effect, although the collapse of the Black Sea Grain Initiative after only one year in 2023 resulted in a surge in wheat prices.

We assessed the trade policy regimes that underpin trade flows with external partners. We saw that many African countries' agricultural exports benefit from market access concessions such as the EU's Everything but Arms, the US's African Growth and Opportunity Act and the WTO's initiative of duty-free, quota-free market access for least-developed countries. But there is a high level of agriculture sector protectionism in bilateral partners' markets through measures allowed by WTO rules. These include high import tariffs for farm products and subsidies to farmers, which lead to both overproduction and enhanced levels of GHG emissions in some food production sectors. Agricultural protectionism makes many African food exports less competitive, especially in traditional partners' markets. Capacity in several African countries to meet food safety standards required for exports is a perennial challenge. In the case of China, significant efforts have been made to work with African exporters to ease this difficulty through the introduction of 'green lanes'. Policies in the EU related to its Green Deal and Fit for 55 such as the Carbon Border Adjustment Mechanism and deforestation regulation will increasingly expose the nexus between trade and climate to greater scrutiny.

10.9 Agricultural negotiations remain contentious at the WTO, with limited progress

Finally in Chapter 9, we turned our attention to the WTO and noted that agricultural negotiations remain contentious, with limited progress in addressing imbalances and asymmetries. We argued that, while it is in the interest of the African group to work towards revitalising multilateral agricultural trade reform, trade-offs are inevitable as some reforms might be desirable in relation to African agricultural production but not in relation to consumption. Others would not only secure policy space and flexibilities for African Members but would simultaneously provide benefits for emerging countries with large agricultural production volumes, like China and India, with potentially negative implications for African agricultural producers.

We suggested that African Members would be advised to be pragmatic in agricultural negotiations, i.e. adopting an approach that focuses on results over principles, evidence and technical analysis over ideological positioning. We further suggested that African Members should call for new research that can offer fresh insights on multilateral trade rules and food security in a changing global economy and in the context of sustainability and the climate emergency.

We noted that most African Members have not used the Development Box for providing domestic support up to their allowed *de minimis* levels. This suggests that the problem is not necessarily a lack of policy space but rather fiscal space along with national policies and priorities.

At the WTO, food security, agriculture and the environment are discussed in different fora and committees. Sustainability issues such as efficiency of water use in agriculture, the safety of food products derived from new technologies, food waste in supply chains from farm to consumer, and knowledge and technological applications for increasing product yield and reducing farm emissions need to be mainstreamed into agricultural negotiations, as advocated by the African Members (World Trade Organization 2023). This calls for enhanced cooperation between these relevant bodies, to streamline the discussions and bridge the silos. With very small delegations in Geneva, African countries will surely benefit from a rationalisation of the food security agenda at the WTO. This is also necessary to understand the trade-offs better and to make sure that effective solutions are reached.

10.10 Final word: the story of food deprivation in Africa is complex but overcoming poverty matters most

Our overall conclusion on why so many Africans endure food deprivation is that the story of Africa's food trade, food security and climate risks is complex, and resists being reduced to a simple, comprehensive narrative. But the keys that will unlock a sustainable pathway to food security on the continent are (1) economic growth and rising incomes that are widely shared; (2) more value added in regional and global value chains to transform the economic model that underpins poverty, unemployment and food deprivation on the continent; (3) attention to policy implementation, finance, investment, institutions, knowledge, capacities and behind-the-border trade measures that are required for boosting agricultural productivity and output and bringing trade costs down; and (4) adaptation and mitigation strategies in response to climate risks that are responsive to fast-changing knowledge and technologies. These are what matters most to bring down the high numbers of malnutrition and hunger among Africans.

This book has unpacked the critical issues that underlie food security in Africa. We have provided insights for interrogating policy choices and fiscal provisioning and for activism and campaigning at various levels of government, economy and society that interface food security. The book is published on an open access basis to make it easily accessible and to enrich discussion and engagement on the issues. Researchers are encouraged to go deeper into the ideas covered in this book. For teachers and educators, the book can be used in interdisciplinary courses on international development and across several disciplines in the social sciences including economics, law, politics and international relations. Most importantly, it is hoped that this book will help to generate a fuller understanding on what needs to be done to overcome food insecurity in Africa.

Notes

[1] Author's analysis based on United Nations Population Division (n.d.).

[2] NB: the Sustainable Development Goals target the total eradication of extreme poverty (defined for the purposes of the goals as persons living on less than $1.25 per day) and a halving of poverty according to national definitions (United Nations n.d.).

References

Fofana, Ismail; Chitiga-Mabugu, Margaret; and Mabugu, Ramos E. (2023) 'Is Africa on Track to Ending Poverty by 2030?', *Journal of African Economies*, vol. 32, Supplement 2, pp.ii87–ii98. https://doi.org/10.1093/jae/ejac043

Mangeni, Francis; and Mold, Andrew (2024) *Borderless Africa: A Sceptic's Guide to the Continental Free Trade Area*. London: Hurst Publishers.

Population – Total (thousands) United Nations Economic Commission for Africa. https://ecastats.uneca.org/data/browsebyIndicator.aspx

United Nations (n.d.) '1 End Poverty in All Its Forms Everywhere', United Nations | Department of Economic and Social Affairs Sustainable Development. https://perma.cc/QQ6F-JTJG

United Nations Economic Commission for Africa (2024) *Economic Report on Africa 2024*, United Nations. https://perma.cc/9A53-B4XE

World Trade Organization (2023) 'Communication from the African Group, Role of Transfer of Technology in Resilience Building: Agriculture', Geneva: WTO.

Index

Page numbers in *italic* or **bold** refer to a figure or table respectively.

UNCTAD (United Nations
    Conference on Trade and
    Development) product group
    classifications, 13
undernourishment, 2
UNICEF (United Nations Children's
    Fund), 2
United Arab Emirates (UAE), trade
    with, *188*, **190**, **191**
United Kingdom (UK), trade with,
    188, *189*, **190**, **191**, **193**,
    194, 196
United Nations (UN) agencies, 2–3,
    4, 68
United Nations Women's
    Entrepreneurship
    Development Programme, 83
United States of America (US, USA)
    exports to, *188*, **190**, **194**
    imports from, 22, *189*, **190**, **191**,
    193, **195**, 196
    trade policies, 200–201, 215
urbanisation, 43

V

value chain financing, 78
vanilla, 17, 19
vegetable products, 128, **129**, **134**
vegetables, 16, 18, **55**, 56, *109*, **132**, **137**
venture capital investment, 78
Viet Nam, trade with, 18, 22

W

WAEMU (West African Economic
    and Monetary Union), *111*
weather extremes, 28
West (Western) Africa
    climate change risks, **55**
    exports, 18, 19, 114
    food insecurity, 2, 24, 25, *26*
    food production and consumption,
        35, *36*, *39*, 40, 41, 42, 44, 46,
        48, 49, 50, 51, *52*, 53, 54
    imports, 22, 23

intra-African trade, potential
    impact of AfCFTA, *142*, 143
    regional trade, 114, 116, 117
WFP (World Food Programme), 2,
    4, 226
wheat
    agricultural subsidies, 218
    bilateral trade, 21, 22, 197
    climate change impacts, 27, 28,
        47, 56
    exports, 16
    imports, 21, 22, *110*, *113*, 116, 197
    intra-African trade, *109*, *110*, *113*,
        116, **118**, **132**, **137**
    production and consumption, 37,
        38, *46*, 46–47, *47*
    tariffs, **132**
    trade diversion, potential, **137**
women in agriculture, 77, 83, 175,
    176, 252
World Agroforestry Centre, 83
World Food Programme (WFP), 2,
    4, 226
World Health Organization
    (WHO), 2
World Trade Organization (WTO),
    4, 213–235, 253–254
    Agreement on Agriculture (AoA),
        215–220, 221, 223, 225
    Aid for Trade initiative, 83
    climate and environmental
        challenges, 231–233
    export restrictions, 224–227
    Fisheries Subsidies Agreement
        (FSA), 228–229
    food security and WTO rules, 214,
        233–235
    Food Security Declaration,
        225–226, 226
    Ministerial Decision on WFP
        Exemptions, 226
    public stockholding programmes,
        221–223
    Sanitary and Phytosanitary
        Agreement, 227

www.ingramcontent.com/pod-product-compliance
Lightning Source LLC
Chambersburg PA
CBHW051954270326
41929CB00015B/2654